T0214692

SpringerBriefs in Climate Studies

SpringerBriefs in Climate Studies present concise summaries of cutting-edge research and practical applications. The series focuses on interdisciplinary aspects of Climate Science, including regional climate, climate monitoring and modeling, palaeoclimatology, as well as vulnerability, mitigation and adaptation to climate change. Featuring compact volumes of 50 to 125 pages (approx. 20,000–70,000 words), the series covers a range of content from professional to academic such as: a timely reports of state-of-the art analytical techniques, literature reviews, in-depth case studies, bridges between new research results, snapshots of hot and/or emerging topics Author Benefits: SpringerBriefs in Climate Studies allow authors to present their ideas and readers to absorb them with minimal time investment. Books in this series will be published as part of Springer's eBook collection, with millions of users worldwide. In addition, Briefs will be available for individual print and electronic purchase. SpringerBriefs books are characterized by fast, global electronic dissemination and standard publishing contracts. Books in the program will benefit from easy-to-use manuscript preparation and formatting guidelines, and expedited production schedules. Both solicited and unsolicited manuscripts are considered for publication in this series. Projects will be submitted to editorial review by editorial advisory boards and/or publishing editors. For a proposal document please contact the Publisher.

More information about this series at http://www.springer.com/series/11581

Nader Noureldeen Mohamed

Energy in Agriculture Under Climate Change

 Springer

Nader Noureldeen Mohamed
Department of Soil and Water Sciences
Faculty of Agriculture
Cairo University
Giza, Egypt

ISSN 2213-784X   ISSN 2213-7858   (electronic)
SpringerBriefs in Climate Studies
ISBN 978-3-030-38009-0   ISBN 978-3-030-38010-6   (eBook)
https://doi.org/10.1007/978-3-030-38010-6

© The Author(s), under exclusive license to Springer Nature Switzerland AG 2020
This work is subject to copyright. All rights are solely and exclusively licensed by the Publisher, whether the whole or part of the material is concerned, specifically the rights of translation, reprinting, reuse of illustrations, recitation, broadcasting, reproduction on microfilms or in any other physical way, and transmission or information storage and retrieval, electronic adaptation, computer software, or by similar or dissimilar methodology now known or hereafter developed.
The use of general descriptive names, registered names, trademarks, service marks, etc. in this publication does not imply, even in the absence of a specific statement, that such names are exempt from the relevant protective laws and regulations and therefore free for general use.
The publisher, the authors and the editors are safe to assume that the advice and information in this book are believed to be true and accurate at the date of publication. Neither the publisher nor the authors or the editors give a warranty, expressed or implied, with respect to the material contained herein or for any errors or omissions that may have been made. The publisher remains neutral with regard to jurisdictional claims in published maps and institutional affiliations.

This Springer imprint is published by the registered company Springer Nature Switzerland AG
The registered company address is: Gewerbestrasse 11, 6330 Cham, Switzerland

Preface

Climate change, water and energy are the major factors controlling the world food production as well as having direct impacts on food security, especially in the developing countries (water, energy food and climate change nexus; and indirectly will include population growth). The nexus of water, energy and food impacts the environment and communities around the world. Agriculture has the largest water footprint of any other activity and consumes a summation of 70% of global water. Agriculture sectors and the food industry consume more than 30% of the total global energy consumption. From the role of water in energy production to the 3000–5000 l of water needed to produce 1 kg of rice, the interconnected systems of water, energy and food present challenges and opportunities of global significance.

Water is the backbone of agriculture production before energy or soil; even water, energy and food become one nexus. It is well known that 15% of the global energy comes from water sources and 17% of energy is consumed by water delivery systems. Energy usage in water includes water collection, processing, distribution and end-use power requirements, such as pumping, transport, treatment and desalination. Water in energy includes thermoelectric cooling, hydropower, mineral extraction, mining, fuel production (fossil, non-fossil and biofuel) and gas emission control.

In the year 2050, the demands for water will increase by 30%, but under global heating the total water resources will decrease by 6%. Thus, food production will decrease by 12–20% and land degradation will increase according to the buildup of soil salinity due to the increase of soil solution evaporation. Increasing soil salinity means more water is needed to leach out salinity from the root zone of the surface soils. The decrease of accessible water will increase the need for treating all types of wastewater, in terms of reuse and desalination, for both sea water and marginal poor quality water that contains a lot of salts and minerals.

The frequencies of droughts and dryness will increase due to global heating, especially in upstream river basins; thus the need for extracting more groundwater even the deepest ones, as well as the need for water desalination, will increase sharply, which in turn will lead to more energy consumption.

The need for agro-food and other agriculture productions will increase by 60% in 2050, which means increase of the need for more chemical fertilizers and more pesticides and this increase of use is considered as a high intensive consumption of energy that leads to more energy consumption. Water scarcity under global heating will increase the need for intensive food production under greenhouse systems in addition to the need for air-conditioning, fast transportation and freights for food marketing which also increases the energy consumption. Increasing the agriculture activity and improving the lifestyle of rural areas means more gas emission and more energy are needed to meet these necessary requirements.

Cairo, Egypt   Nader Noureldeen Mohamed
August 2019

Introduction

Climate change has direct effects on all agriculture resources, such as water resources, wetlands, fresh water ecosystems, land degradation, food and other agriculture production, forestry, biodiversity and contamination. Climate change also affects agriculture indirectly through its impacts on energy, industry, commerce, financial service, human settlement, health, coastal zone, and marine and low deltas ecosystems (International Journal of Humanities and Social Science [December 2011], Edame et al. 2011). On the other hand, the 2010 World Development Report draws on the analysis of the Intergovernmental Panel on Climate Change (IPCC 2007a) to calculate that agriculture directly accounts for 14% of global GHG emissions in CO_2 and equals and indirectly accounts for an additional 17% of emissions when land use and conversion for crops and pasture are included in the calculations, with a total of 31%. Agriculture's share in global GDP is only 4%, which means that agriculture is a high greenhouse gas intensive producer (International Journal of Humanities and Social Science [December 2011]; IPCC 2007a; Edame et al. 2011; IPCC 2007b; NASA 2011).

Agricultural production directly relies mostly on weather conditions that include water from rain, surface and ground, which—together with soil conditions—controls the degree of plant growth and field productions. In some cases, weather conditions can be managed by using irrigation to replace the deficient rainfall or shifting the timing of the cropping season and replacing the weather humidity to avoid adverse weather conditions (e.g., irrigated rice and sugarcane in Egypt instead of rain-fed agriculture in some regions). Cultivation under greenhouse systems is another solution for tough weather conditions and water shortages, which provides completely controlled environments and precision in temperature, radiation and all other inputs. Moreover, and at the same time, it is considered as economically feasible in small-scale cases and for high-value crops. Some cases of extreme weather that lead to severe damage cannot be managed, such as strong winds, floods, hail or frost and drought (United Nations Development Programme (UNDP) 2007; FAO 2016; Food and Agriculture Organization of the United Nations (FAO) 2000).

FAO (2016) reported the effects of climate change on agriculture and food security, which stated that:

1. Climate change already negatively impacts agriculture and food security; thus, without urgent action, this impact will put millions of people at risk of hunger and poverty.
2. The impacts on agricultural yields and livelihoods will vary across countries and regions, but it will become increasingly adverse over time and potentially catastrophic in some areas.
3. The world should work hard to limit global temperature increases to 1.5 °C above pre-industrial levels, which would significantly reduce the risks and impacts of climate change.
4. Deep transformation in agriculture and food systems from pre-production to consumption is needed in order to maximize the co-benefits of climate change adaptation and mitigation efforts.
5. The agriculture sectors have the potential to limit their greenhouse gas emissions (31% of the total global emission), but at the same time, ensuring future food security requires a primary focus on adaptation.

Climate change has already become a problem that should be adapted to and mitigated. It is seriously affecting the agricultural productivity and food security even in developed or developing economies. According to (Barghouti 2009) the decline in expected food production due to climate change can be avoided through raising the irrigation efficiency, watershed management, good soil and water management, improving cultivation systems and livestock management in addition to the use of high productive seed and using plant breeding to adapt plants to frequent dryness, salinity and heating under higher temperature. Plant and animal biodiversity should increase resilience and adaptation to changing environmental conditions and stress, such as drought, salts, contaminators accumulation and flooding as an extreme event. Land use for livestock production, including grazing lands dedicated to the production of feed, represents approximately 70% of all agricultural land in the world (Barghouti 2009).

Thus, many developing countries consider adaptation the main priority because of the significant impacts of climate change which is expected to have on national development, sustainability and national security.

Climate Change, Water, Energy and Food Security

Climate change, water and energy are the major factors controlling the world food production and have direct impacts on the food security, especially in the developing countries (water, energy, food and climate change nexus; and should include population growth). The nexus of water, energy and food impacts the environment and communities around the world. Agriculture is the largest water consumer of any human activity and represents 70% of global water withdrawal. Meanwhile, agriculture and the food industry accounts for 30% of global energy consumption. From the role of water in energy production to the 3000–5000 l of water needed to produce 1 kg of rice, the interconnected systems of water, energy and food present challenges and opportunities of global significance.

Climate change has become a real fact and a phenomenon that affects global heating, regional rain patterns and biological systems. Attributed directly or indirectly to human activity, climate change puts additional pressure on already over-exploited natural resources. It negatively affects crop yields, stability of food supplies and the ability of people to access and utilize food in many parts of the developing world. Although rich economic countries are responsible for most greenhouse gas emissions (GHGs), the impact of climate change is expected to be the most severe in developing countries. Low-income communities rely mostly on agriculture, forestry and fisheries, thus they are highly climate-sensitive and more vulnerable to losing their little resources. A study by (Mohcine Bakhat and Würzburg 2013; Food and Agriculture Organization of the United Nations (FAO) 2000) stated that the risks the climate change poses on food security are particularly pressing at a time of high oil prices, at levels surpassing $130 a barrel in May 2008. High fuel prices make agricultural production more expensive by raising the cost of producing fertilizers, irrigation motors pumps, machines operation, shipping freights (tractors, planting and harvesting machine etc.), lighting and transportation from fields to markets. With high oil prices up to year 2010, farmers have switched massively to production of crops for ethanol and biodiesel, but with decreasing prices after 2012 they returned back to produce food crops. The increased level and volatility of agricultural prices is negatively impacting the purchasing power and the food security of the poor (Mohcine Bakhat and Würzburg 2013). Figures 1 and 2 show the relation between oil and food price and the bioethanol production to replace fossil fuels.

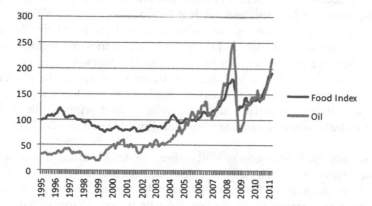

Fig. 1 The relation between oil and food prices (Mohcine Bakhat and Würzburg 2013)

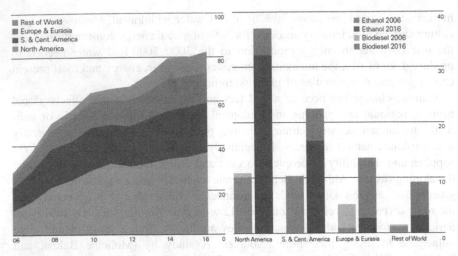

Fig. 2 Bioethanol and biodiesel production up to 2016 (Food and Agriculture Organization of the United Nations (FAO) 2000)

Food security according to FAO 2000 (Food and Agriculture Organization of the United Nations (FAO) 2000) is not only the supplied sufficient quantities and quality food through domestic production or imports, but it also depends upon the surrounding environment and community condition, such as the availability or the possibility to access this food. Food should not be expensive to avoid "hide hunger" according to food soaring prices and should be accessible for the poor and wealthy according to the UN principal of rights "Right to Food": extreme cold or warmth, conflicts or civil war, insecure condition and so on. Food should also be healthy and not pathogenic or carcinogenic. Household food access is the ability to acquire sufficient quantities of quality food to meet the needs of all household members' "nutritional requirements". Access to food is determined by physical and financial resources, as well as by social and political factors.

The effects of climate change on agricultural production and livelihoods are expected to intensify over time, and to vary across countries and regions. Beyond 2030, the negative impacts of climate change on the productivity of crops, livestock, fisheries and forestry will become increasingly severe in all regions. Productivity declines would have serious implications on food security. Food supply shortfalls would lead to major increases in food prices, while increased climate variability would accentuate price volatility (Pinterest site 2012; Barghouti 2009).

On the contrary, food insecurity exists when people lack sustainable physical or economic access to enough safe, nutritious and socially acceptable food for a healthy and productive life. Food insecurity may be chronic, seasonal or temporary. Nutritional consequences of insufficient food or undernutrition include protein energy malnutrition, anemia, vitamin A deficiency, iodine deficiency and iron deficiency.

Food insecurity may also result in severe social, psychological and behavioral consequences. Food-insecure individuals may manifest feelings of alienation, powerlessness, stress and anxiety, and they may experience reduced productivity, reduced work power ability, school performance and reduced income earnings. Household dynamics may become disrupted because of a pre-occupation with obtaining food which may lead to indignation or anger, pessimism and irritability. Adverse consequences for children include higher levels of aggression or destructive behavior, hyperactivity, anxiety, difficulty with social interactions (e.g., more withdrawn or socially disruptive), increased passiveness, poorer overall school performance, increased school absences and a greater need for mental health care services.

Contents

Chapter 1
Climate Change and Agriculture

The relation between climate, water, soil and crop production is so complicated and looks like a circle without a beginning or an end, whereas each item affects all others. Global heating will increase evaporation from all waterbodies (rivers, irrigation canals, lakes etc.) causing a lack of water, especially in hyper arid, arid and semi-arid regions. At the same time, it will decrease the amount of little rainfall in these areas. Global heating will also increase the evaporation and evapotranspiration from irrigated lands, through grasses, weeds and growing plants. Thus, the plants will need more water and more frequent irrigation, and the soil also will need more water due to the salinity building up and salt accumulation. But actually there is a lack of water beside the deterioration of water quality due to the increase of pollutants concentration due to high temperature effects, which will also increase land degradation and lead to the need for more water fraction (as a part of irrigation dose) or water requirement for reclamation of deteriorated lands at the time of water scarcity. Water scarcity in hyper arid and arid countries will lead to an expansion in the reuse of wastewater such as agriculture drainage water which has appreciable amounts of salts, residual chemical fertilizers and residual pesticides which are classified as problematic poor quality water and will cause more land degradation. Thus this poor quality water will need using more water for leaching as leaching fraction or leaching requirements, but the sufficient water to meet all these demands is completely absent. Finally, climate change will decrease the available water and will increase land degradation and the buildup of salt-affected soils and contaminated soils, and this will decrease land productivity which needs sufficient water to leach out the salinity and contaminators, but there is not enough water to meet these demands! The following drawing diagram summarizes the relation between global heating and water resources.

© The Author(s), under exclusive license to Springer Nature Switzerland AG 2020
N. Noureldeen Mohamed, *Energy in Agriculture Under Climate Change*,
SpringerBriefs in Climate Studies,
https://doi.org/10.1007/978-3-030-38010-6_1

1.1 Global Heating Effects

Thus, global heating will increase the following (Barghouti 2009; Müller et al. 2009)

1. Evaporation and evapotranspiration
2. Irrigation water demands
3. Soil salinity and land degradation
4. Water contamination
5. The need for leaching fraction during plant growth
6. Domestic and municipal water demands
7. Pests, plant diseases
8. Aquatic wrack, weeds and grass in irrigation canals

   And will decrease:

1. Availability of water
2. Water quality
3. Yield crop and building up of meat in livestock
4. Water quota for agriculture sector
5. Available water for leaching salinity
6. Precipitation and rain-fed agriculture yield crop
7. Natural pasture and forage areas
8. Profits from agriculture activity.

   Figure 1.1 shows the relation between global warming and water resources.

1.2 Climate Change and Water Resources

Effect of climate change on Water Resources

1.3 Soluble Oxygen

- Temperature: Impacts of global climate change on temperature are perhaps the most obvious ones and are particularly important because temperature is a driver of many other hydrological variables.
- The increase in air temperature will cause an increase in water temperature. As water temperature increases, water contamination problems will increase and the pollutants will be more active and more concentrated; thus, many aquatic habitats will be negatively affected. Generally, the increase in water temperature may result in the following (Irfan 2019):

  – Lowering levels of dissolved oxygen in water due to the inverse relationship between dissolved oxygen and temperature (Fig. 1.2).

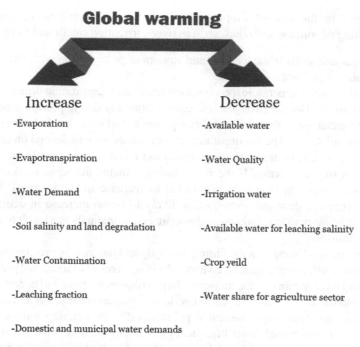

Fig. 1.1   Global heating and water resources (design by Author)

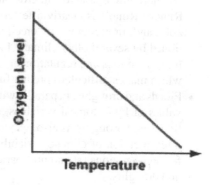

Fig. 1.2   The relation between temperature and oxygen level in water (WATERSHED ACADEMY WEB and US Environmental Protection Agency 2013)

– Increase the activity of pathogens microbe, trace element, nutrients concentration and invasive species.
– Increase the concentrations of some pollutants such as ammonia and pentachlorophenol due to their chemical response to warmer temperatures.
– Increase algal blooms.
– Loss of some aquatic species (weeds and aquatic grasses) which are classified as temperature dependent.
– Change in the abundance and spatial distribution of coastal and marine species and the decline in populations of some species.

- Increase in the rate of evapotranspiration from waterbodies, resulting in shrinking of some waterbodies, such as rivers, irrigation canals and fresh lakes.

- Changes and shifts in the location and amount of precipitation will affect water availability and water quality.
- Soil moisture: Temperature, precipitation and evapotranspiration directly affect soil moisture. But the strongest influence is normally due to precipitation. Soil moisture changes influence strongly crop growth and water needs for irrigation.
- Water availability: The net impact on water availability will depend on changes in precipitation. In areas where precipitation increases sufficiently, net water supplies might increase. If the precipitation remains the same or decreases, net water supplies would decrease (due to increase in evaporation). Where water supplies decrease, there is also likely to be an increase in demand in all sectors (agriculture, industrial, domestic and municipal) as a result of higher temperatures.
- Water quality: Changes in the timing, intensity and duration of precipitation can negatively affect water quality. Extreme flooding, a result of increased precipitation and intense rain storms, transports large volumes of water and contaminants into waterbodies. Flooding can also overload storm, combine sewer and wastewater systems, resulting in untreated pollutants directly entering waterways. In regions with increased rainfall frequency and intensity, more pollution and sedimentation might be produced because of runoff. Reduced rainfall can also result in more frequent wildfires, and land areas where wildfires have occurred are more vulnerable to soil erosion.
- Runoff: Runoff is clearly affected by the above-mentioned hydrological variables and, in particular, by precipitation. However, future runoff is also conditioned by several other climatic factors and human influences, such as streamflow diversions and regulation or interaction between surface and groundwater, which makes difficult to predict future runoff.
- Floods and droughts: In parallel with the impact of climate change on the average values of hydrological variables, the impact on extreme phenomena, such as floods and droughts, is also relevant. Several studies indicate a tendency for an intensification of climate variability in situations of climate change and offer, for some regions, apparently paradoxical scenarios of increase in both floods and droughts.
- Water access will affect:
  1. Reduced ground and surface water supply in arid regions
  2. Increased water demand due to high temperatures
- Water quality will affect:

1. Increased runoff resulting in land erosion and sedimentation
2. Overwhelmed water infrastructure due to flooding.

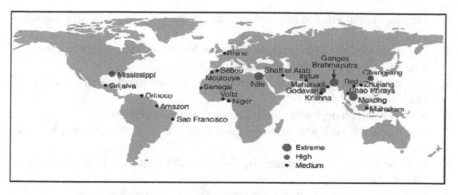

Fig. 1.3 Threatening deltas in case of sea-level rise (IPCC 2007)

1.4 Climate Change and Sea-Level Rise

The factors driving sea-level rise include the following:

1. Ocean water expansion caused by warmer ocean temperatures
2. Mountain glaciers and ice caps melting
3. Greenland Ice Sheet and the West Antarctic Ice Sheet melting
4. Ice cap shrinking and rising sea levels by 0.18–0.59 m (IPCC 2007).

Figure 1.3 shows the threatening deltas in the world.

1.5 Climate Change and Precipitation Changes

As air temperature increases, the rate at which water evaporates from soils and waterbodies increases, and that increases the humidity being held in the atmosphere. Because there is more atmospheric moisture, there are heavier downpours when it rains. While moderate increase in annual average precipitation is expected, there is likely to be a wider variation in the pattern of rainfall, specifically, drier dry periods punctuated by more intense rainfall.

1.6 Climate Change and Agriculture Productions

The impacts of climate changes in agriculture productions could be summarized as (IPCC 2007; https://www.pinterest.co.uk/pin/717901996820657072/):

1. Agriculture sector is one of the most sensitive sectors to climate change
2. Global agricultural productivity could decline between 10 and 25% by 2080

3. Decline in crop yield in rain-fed agriculture can be as much as 50% for some countries
4. Biggest losers are winter crops such as strawberry (−32% in area of cultivation), wheat (−18%), rye (−16%) and oats (−12%)
5. Among the winners are pearl millet (+31%), sunflower (+18%), chickpea (+15%) and soybean (+14%).
6. MENA has high dependency on climate-sensitive agriculture
7. Crops and livestock will face increased heat stress—yield reduction by up to 10%
8. Increase in crop water demand by 5–8% by 2070 with fall in productivity of unit of water
9. Intensified evaporation will increase the salt accumulation in soils
10. Increase in pest infestations in warmer climates will expand the uses of chemical pesticides
11. Rise of sea level poses a threat to agriculture in low-lying coastal areas—particularly the Nile Delta in Egypt
12. Natural ecosystems such as rangelands and forests are less resilient and more vulnerable.

Figure 1.4 shows the decrease in agricultural yield crops in the world by 2050.

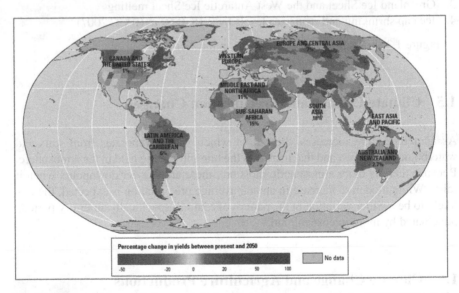

Fig. 1.4 Climate change will depress agricultural yield in most countries by 2050 (Müller et al. 2009; World Bank 2008)

Chapter 2
Energy and Climate Change

2.1 History of Agriculture and Energy

Historically, the agriculture sector was the first source of energy in the world even for heating, cooking and lighting. At first, foster trees were used as sources of wood and other trees residues were used in all home demands. The animal wastes after drying were used in cooking as a source of heat that does not produce smoke, which became later on the first inspiration for the idea of using natural gas, biogas and pressed potages. Animal wastes were also the first form of fertilizers used specially as organic fertilizers before the industrial age and the beginning of producing chemical fertilizers which are considered as intensive energy users, as shown in Fig. 2.1. The idea of using animals and poultry manure as complex nutrients fertilizers, mainly as nitrogen fertilizers because there is no source of it in the soil or in the rocks and minerals which weathered to form soil, can contain or produce nitrogen to release it. Thus, this explains the reason for the obligation of adding nitrogen or organic

Fig. 2.1 The cattle backyards and the rural house oven using primary and dry agriculture wastes (photo taken by the author)

© The Author(s), under exclusive license to Springer Nature Switzerland AG 2020
N. Noureldeen Mohamed, *Energy in Agriculture Under Climate Change*,
SpringerBriefs in Climate Studies,
https://doi.org/10.1007/978-3-030-38010-6_2

7

fertilizers (as a source of nitrogen) before planting all kinds of crops. In the past, the women in Africa and Asia may quarrel, spar and even fight among each other to obtain the fresh manure when they follow after cow or buffalo. Farmers used to add soil particles (soil dusts and mud) that become over the fields sides after the tillage and leveling of fields as a thick layer underneath their livestock in the stockyard to adsorb both solid and liquid wastes of these animals and keep it working until the beginning of the next planting season to cut it and add it to the fields as a, rich in nutrients, natural organic fertilizer.

In this trend, the call for reducing the use of nitrogen fertilizers to the soil is completely wrong, because this is an outer nutrient source only. However, the matter that should be concerning is how we can control the nitrogen losses from the soil even from the denitrification processes as volatile nitrogen oxides; or as easy leaching out for the soluble nitrogen forms in the soils, especially the nitrate form.

Global heating and the frequency of extreme events as a result of climate change such as hurricanes, rivers and flash floods will surely affect the energy consumption. There is also the increasing demand of energy in all faces, electricity or fuels, according to population growth and economic growth.

The energy systems have been developed to meet the changes in environment and climate change.

The impact of climate change on the energy sector is increasing with the growth of climate change, especially greenhouse gas emissions, where 61% of it is produced by the energy in industrial sectors. Moreover, the development of the energy sector depends on the integration between energy, environment and climate policies (Bundschuh 2014; International Energy Agency (IEA) 2014).

2.2 Importance of Energy Under Global Heating

According to the sustainable development goals and the nexus of water, energy, food and climate change, the importance of energy to the economic and its socio-economic impacts can be summarized as follows (Bundschuh 2014; International Energy Agency (IEA) 2014):

1. **Energy is the key factor in economic and social progress**

For sure, no economic progress can be achieved without stable and continuous energy services. Worldwide, approximately two billion people don't have access to modern energy.

2. **Energy has a major impact on the global and local environment**

The UNFCCC has focused on the adverse impacts of energy supply on the environment. The needs of clean energy technologies are recommended.

3. **Insufficient energy for agriculture affects food production**

A wide range of modern and traditional energies forms are used directly on the agricultural farms, such as tractors, water pumping, irrigation and crop drying. The energy is also used indirectly for manufacturing fertilizers and pesticides. Other energy inputs are required for post-harvest processing, packaging, storage and transportation.

4. **There is a general lack of rural energy in agriculture sector**

Developing countries are concerned mainly by industry extensions on their energy policies followed by transportation and urban infrastructures. Agriculture usually contributes to the economy of developing countries by around 11–30% of a country's GDP. Energy for agriculture needs to have a priority in the rural policy and technology assessment in developing countries.

5. **An energy transition is needed in rural areas**

Agriculture can have a major role in supporting rural livelihoods through increase of sourced bioenergy. This can assist rural development and improves food productivity.

6. **The energy function of agriculture should be exploited**

The role of agriculture and agro-industry as an energy supply resource is an important factor in taking forward this energy transition to achieve high sustainable development in the rural areas. Agriculture can also make a major contribution to climate change mitigation by reducing CO_2 emissions since biomass is a high carbon energy source. Agriculture's contribution as a carbon energy resource should be exploited as one solution with other greenhouse gas emission control technologies.

7. **Advancing modern bioenergy technology**

Large-scale production of bioenergy should be developed to use agricultural and forest residues to dedicated energy crop plantations.

2.3 A Challenge and an Opportunity

An integrated role of agriculture as dual function of agriculture as an energy user and an energy supplier needs to be improved. In order to mobilize the synergies and develop the energy function of agriculture (Bundschuh 2014; International Energy Agency (IEA) 2014):

1. The role of agriculture in providing renewable energy should be recognized in future energy policy technology development.
2. Research related to potential to deliver the energy function of agriculture in a cost-effective and market-oriented manner is needed.
3. There is a need for positive political support to make social and cultural changes that might be needed to develop this potential.

2.4   Facts on Energy and Climate Change

A recent study on energy and climate change has been published from The Hamilton
Project and Energy Policy Institute at the University of Chicago (2017), summarizing
the economic facts on energy and climate change as follows:

1. Global temperature and heat is rising and will continue to be raising more
    (Fig. 2.2).
2. Global temperatures will keep rising by another 8 °C, under the current rate of
    using global fossil fuel, or more.
3. The cumulative emissions from China and India will continue to grow by 2100
    (Fig. 2.3).
4. Before the end of this century, all US states are projected to have a temperature
    increase.
5. The infrastructure in the USA is vulnerable against the impacts of climate
    change.
6. The social costs will be higher than the prices we pay for fossil fuels.
7. With the impacts of climate changes, USA has to depend on more renewable
    sources instead of coal and other fossil fuels before 2050.
8. The total emission of transportation from carbon now exceeds electric power
    emissions (Fig. 2.4).
9. Total emission from vehicles that are powered by natural gas is less than that
    from coal plants.

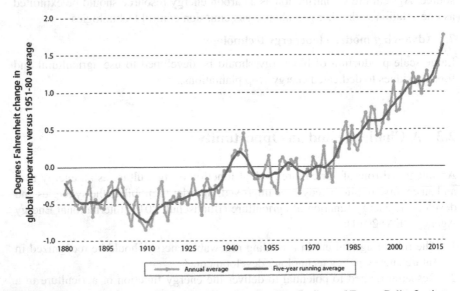

Fig. 2.2 Global temperature change, 1880–2015 (The Hamilton Project and Energy Policy Institute
at the University of Chicago 2017)

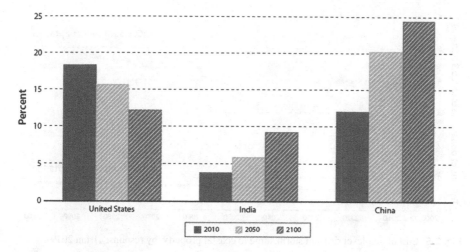

Fig. 2.3 Share of cumulative GHG, 2010–2100 (projected) (International Energy Agency (IEA) 2014)

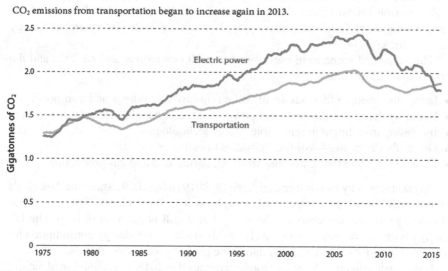

Fig. 2.4 US CO_2 emissions by sectors, 1975–2016 (Energy Information Administration (EIA) 2016)

10. Investment in energy sector should be increased because it still remains below 1970s and 1980s levels.
11. To lower the burden of climate change, investment in climate adaptation should be greater (Fig. 2.5).

Climate change mitigation and modification in energy sector relies on increasing new investments in clean energy and controlling the impacts of high emissions. The

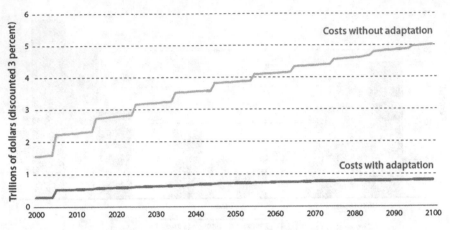

Fig. 2.5 Cost of sea-level rise and storm surge to coastal property, by response (Irfan 2019)

main target in energy is reducing the temperature rise below 2 °C. High carbon taxes could drive changes in infrastructure and enhance policy options toward unlocking high-emission infrastructure. The coal plants, which are considered as one of the largest sources of GHG emissions, provide useful insights (International Energy Agency (IEA) 2014).

The suggested scenario in energy-related GHG emissions around 2020 and the proposed measures are:

- Increasing energy efficiency in all sectors (industry, buildings and transport).
- Reducing the use of the coal-fired power plants and banning their construction.
- Increasing investment in renewable energy technologies.
- Gradually deducting fossil fuel subsidies to end-users by 2030.
- Reducing the methane emissions from oil, agriculture and gas production.

A recent report by Umair Irfan, on April 18, 2019 (Irfan 2019), states the America's record high energy consumption, and explains it in three charts. The report also showed that renewable energy is booming, but it still needs a lot of help. The US Energy Information Administration (Irfan 2019) added: US energy consumption hit a record high in 2018 in large part due to the growing use of fossil fuels.

Fossil fuels still provide 80% of total energy used in 2018. Consumption of natural gas and petroleum grew by 4%, while coal consumption declined by 4% compared to the year before. Renewable energy production also reached a record high last year, climbing 3% relative to 2017, as shown in Fig. 2.6.

The growth in energy uses is largely a function of the growing economy. More goods production and consumption, more transportation and more services mean more energy needed. However, that also means we are moving away from fighting climate change.

The EIA's latest numbers revealed how far we have to go if we want to control global warming target at 1.5 °C above pre-industrial levels this century.

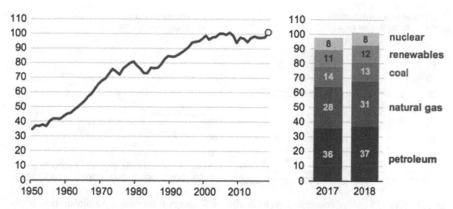

Fig. 2.6 US total energy consumption (1950–2018), quadrillion British thermal units (Irfan 2019; US Energy Information Administration 2019)

The same trend was observed in the report from the Rhodium Group this year, which recorded that the US greenhouse gas emissions rose 3.4% in 2018 compared to the year before, after a three-year decline in the production of heat-trapping gases that contribute to global heating.

2.5 The Six Key Facts from the EIA Report Are

(1) Improving energy efficiency, controlling global heating.

Even the year 2018 reflects a record high in US energy use, but it does not exceed 0.3% higher than the previous recorded in 2007.

Higher demand in 2018 was related to extension uses in natural gas this year during the very cold winter. EIA previously noted that while energy production in the USA is poised to grow, the overall energy use is likely to minimize the changes in the coming decades as energy efficiency improves, as shown by Fig. 2.7.

(2) Reducing emissions is not impossible, but needs much works

The United States has made remarkable progress in cutting emissions from electricity, largely by replacing natural gas instead of coal in electricity generation. Natural gas produces about half of the carbon dioxide emissions that coal releases to produce the same amount of electricity.

But as of 2018, the numbers show that progress is vulnerable to things like the weather, and while natural gas is cleaner than coal, it still has an appreciable value of carbon footprint.

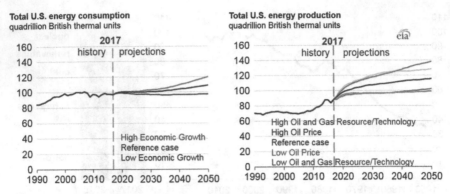

Fig. 2.7 Total US energy consumption (Irfan 2019; US Energy Information Administration 2019)

(3)   Renewable energy is booming, but it still needs help

Wind and solar become the largest sources of electricity generation in the USA as hardware and installation costs continue to decline. However, renewables still have a little share of the overall energy mix.

If the goal is to decarbonize the economy, technology improvements and market forces won't be enough to take a bite out of greenhouse gas emissions. Instead, the USA needs more states and cities—not to mention the federal government—to mandate a full transition to clean energy.

(4)   Electricity generation is not the only gas emission problem

While the electricity generation emissions have declined, the transportation emissions have grown. Transportation has become the largest source of greenhouse gas emissions in the USA. Even though cars and trucks are becoming more fuel efficient and increasingly running on electricity, other modes of transit, namely air travel, still run on fossil fuels. Growing demand for flights is a key reason why US energy use and emissions rose in 2018 (Fig. 2.8).

(5)   Coal in America is reducing

Coal was the oldest and only fossil fuel that is still used in different branches of energy sector. Coal use has declined in 2018. The uses of coal continued to decrease during 2018. The US retired 16 GW of coal-fired power capacity and saw coal consumption dip to its lowest levels in 39 years in 2018.

(6)   Continuous uses of fossil fuels mean more damage to the global heating

As greenhouse gas emissions continue to rise, scientists give more warning that the world is heading toward more global heating. Delaying action means more hazards and more disasters (Figs. 2.9, 2.10, 2.11, and 2.12).

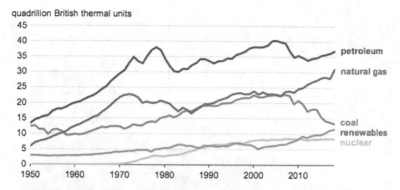

Fig. 2.8 Total energy US consumption, 1950–2015 (Irfan 2019; US Energy Information Administration 2019)

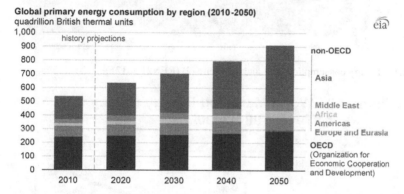

Fig. 2.9 Energy consumption, past and the future (U.S. Energy Information Administration (EIA) 2017)

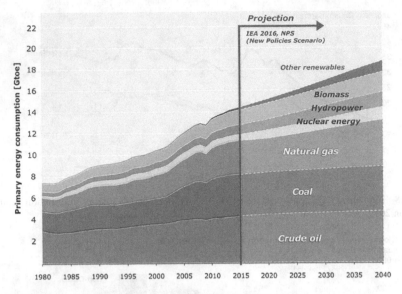

Fig. 2.10 Energy consumption, past and the future (2) (REN21 2012)

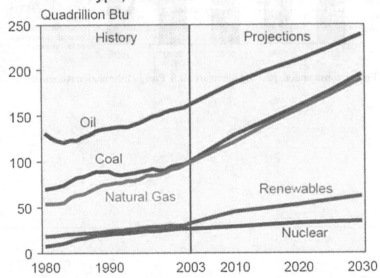

Sources: **History:** Energy Information Administration (EIA), *International Energy Annual 2003* (May-July 2005), web site www.eia.doe.gov/iea/. **Projections:** EIA, System for the Analysis of Global Energy Markets (2006).

Fig. 2.11 World CO_2 emissions by fuel types (FAO 2006)

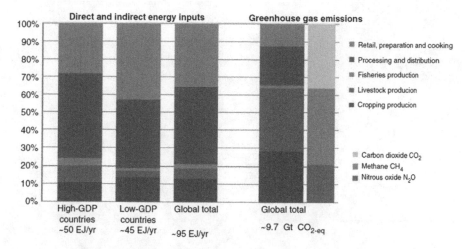

Fig. 2.12 Global sectorial shares on energy consumption of the total food sector and related GHG emission and distribution to high and low GDP countries (FAO 2012; Food and Agriculture Organization of the United Nations (FAO) 2011a)

Chapter 3
Agriculture in Energy

3.1 Introduction

Historically, agriculture is the first source for energy and lighting. First generation of human beings relied on forest woods, pellets and forest residues in addition to the mountain trees, shrubberies and dry flora for cooking, heating, lighting, and to pushing out or getting rid of the insects as well as scaring away savage animals. With agriculture development, people would use vegetable residues, legume stakes, rice and wheat straw, molasses and other sugar cane residues, maize dry plants and corn cob. Then they would use extreme dry animal stools (in cooking because it produces little smolder) and livestock fats with oil seed in lighting.

All these types of energy mentioned have great side effects and impacts the health of people, especially their aspiratory systems due to heavy smokes (smolder) produced from these raw primary energy materials.

Recently, with the progress of education and recent technologies, the world is changing to produce modern and clean energy such as biogas which is produced from fermentation of plants and livestock's residues to use in the rural area for cooking and to improve people's health.

3.2 Biofuel Production Age

Thirty or forty years ago, Brazil led the world in producing biofuel to break free from the Arab bondage as the main producer and seller of the petroleum oil as well as controlling almost all oil trades. Brazil produced bioethanol from sugarcane (as a simple mono sugar able to be fermented to produce bioethanol) and then from starch (as complex polysaccharides sugar) crops depending mainly on their abundant production of sugar, as the highest first sugar producer in the world, to replace fossil gasoline. Brazil continued to develop their production, followed by USA and China

© The Author(s), under exclusive license to Springer Nature Switzerland AG 2020
N. Noureldeen Mohamed, *Energy in Agriculture Under Climate Change*,
SpringerBriefs in Climate Studies,
https://doi.org/10.1007/978-3-030-38010-6_3

to produce bioethanol from maize and then continuing later on to make progress in producing biodiesel from all oil crops to replace fossil diesel in the process. European countries produce bioethanol from sugar beet, wheat, barley, potato, some sweet fruits like water melons (especially in the USA) and the residues of bread and pizza from restaurants and hotels. Asian and African countries use palm dates oil, coconuts, cotton seeds oil and castor oil. Second-generation biofuel relies mostly on plant and forest residues instead of plant crops to produce different types of bioenergy to avoid the competition between energy and food. The following figures explain the development of producing energy from agriculture (from 3.1, 3.2, 3.3, 3.4 and 3.5).

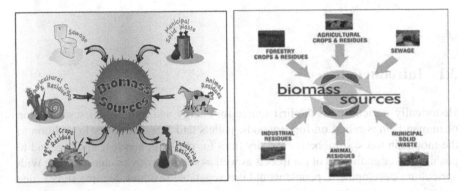

Fig. 3.1 Biomass sources (Food and Agriculture Organization of the United Nations (FAO) 2011a)

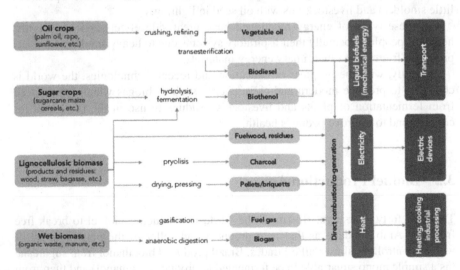

Fig. 3.2 Biofuel productions—input and output (http://www.fao.org/docrep/014/i2454e/i2454e00.pdf; GEMCO ENERGY 2011)

Fig. 3.3 Biofuel raw materials—first and second generation (http://www.fao.org/docrep/014/i2454e/i2454e00.pdf; GEMCO ENERGY 2011)

Fig. 3.4 Biomass, bullet and plant residues (http://www.fao.org/docrep/014/i2454e/i2454e00.pdf; GEMCO ENERGY 2011)

3.3 Some Definitions Related to Bioenergy (Food and Agriculture Organization (FAO) 2000a)

Bioenergy
Bioenergy is defined as the fuel produced from different kinds of agricultural production. It does not include human or animal work. It covers all energy forms derived

Fig. 3.5 Global bioenergy
production (Steenblik 2007;
World Energy Council 2016)

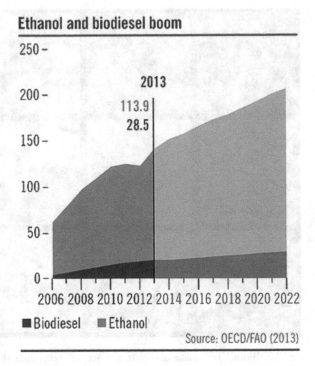

from organic of biological origin. It comprises both purposes, as energy crops and
by-products (residues and wastes). The term by-products includes solid, liquid and
gaseous derived from human activities.

Biomass
Biomass is defined as all forms of plant-derived matter other than that has been
fossilized.

Biofuels: Organic primary and/or secondary fuels derived from biomass can be used
for the generation of thermal energy by combustion or using other technologies. They
comprise both purposes grown energy crops, as well as multipurpose plantations and
by-products (residues and wastes). The term "by-products" includes the improperly
called solid, liquid and gaseous residues and wastes derived from biomass processing activities. There are three main biofuel categories: *woodfuels, agro-fuels and
municipal wastes.*

Woodfuels: This includes all types of biofuels derived directly and indirectly from
trees and shrubs grown in forest and non-forest lands. Woodfuels include biomass
derived from silvicultural activities (thinning, pruning etc.) and harvesting and log-
ging (tops, roots, branches etc.), as well as industrial by-products derived from primary and secondary forest industries which are used as fuel. They also include woodfuels derived from *ad hoc* forest energy plantations. Taking into account the available

database, it is interesting to classify woodfuels into four groups: direct woodfuels, indirect woodfuels, recovered woodfuels and wood-derived fuels, defined as follows.

Direct woodfuels: consist of wood directly removed from **forests** (*natural forests and plantations*; land with tree crown cover of more than 10% and area of more than 0.5 ha).

Other wooded lands: (land either with a tree crown cover of 5–10% of trees able to reach a height of at least 5 m at maturity in situ; or crown cover of more than 10% of trees not able to reach a height of 5 m at maturity in situ, and shrub or bush cover); and **other lands** to supply energy demands and include both inventoried (recorded in official statistics) and non-inventoried woodfuels. Direct woodfuels can be divided into primary and secondary fuels, depending on whether they are directly burned or are converted into another fuel, such as charcoal, pyrolysis gases, pellets, ethanol and methanol.

Indirect woodfuels: usually consist of industrial by-products, derived from primary (sawmills, particle boards, pulp and paper mills) and secondary (joinery, carpentry) wood industries, such as sawmill rejects, slabs, edging and trimmings, sawdust, shavings and chips bark. They preserve essentially the original structure of wood and can be used either directly or after some conversion to another biofuel.

Recovered woodfuels: refer to woody biomass derived from all economic and social activities outside the forest sector, usually wastes from construction sites, demolition of buildings, pallets, wooden containers and boxes, and so on, burned as they are transformed into chips, pellets, briquettes, powder, and so on.

Wood-derived fuels: refer to woodfuels produced in the forest sector, which require several thermochemical processes before use. They do not preserve any trace of the original wood's physical structure, as is the case with black liquor and methanol produced from wood.

Black liquor: is the alkaline-spent liquor obtained from the digesters in the production of sulfate or soda pulp during the process of paper production, in which the energy content is mainly derived from the content of lignin removed from the wood in the pulping process.

Agro-fuels: are obtained as a product of agriculture biomass and by-products. It covers mainly biomass materials derived directly from *fuel crops* and *agricultural, agro-industrial and animal by-products*.

Fuel crops: are employed to describe species of plants cultivated on fuel plantations or farms to produce raw material for the production of biofuel. The fuel crops can be produced on land farms (manioc, sugarcane, euphorbia, etc.), on marine farms (*algae*) or in fresh water farms (*water hyacinths*). The land-produced fuel crops can also be classified under *sugar/starch crops, oil crops* and *other energy crops*. **Sugar/starch crops**: are crops planted basically for the production of ethanol (ethyl alcohol) as a fuel mainly used in transport (on its own or blended with gasoline). Ethanol can be produced by the fermentation of glucose derived from sugar-bearing plants (like sugarcane) or starchy materials after hydrolysis. **Oil crops** cover oleaginous plants (like sunflower, rape, etc.) planted for direct energy use of vegetable oil extracted, or as raw material for further conversion into a diesel substitute, using trans-esterification processes. **Other energy crops** include plants and specialized crops more recently considered for energy use, such as elephant grass (*Miscanthus*), cordgrass and galinggale (*Spartina* spp and *Cyperus longus*), giant reed (*Arundo donax*) and reed canary grass (*Phalaris arundinacea*).

Agricultural by-products: are mainly vegetal materials and by-products derived from production, harvesting, transportation and processing in farming areas. It includes among others, maize cobs and stalks, wheat stalks and husks, groundnut husks, cotton stalks, mustard stalks, and so on.

Agro-industrial by-products: refer to food processing by-products, such as sugarcane bagasse, rice/paddy husks and hulls, coconut shells, husks, fiber and pith, groundnut shells, olive pressing wastes, and so on.

Animal by-products: refer to dung and other excreta from cattle, horses, pigs, poultry and, in principle, humans. It can be dried and used directly as a fuel or converted to biogas by fermentation. **Biogas** is a by-product of the anaerobic fermentation of biomass, principally animal wastes by bacteria. It consists mainly of methane gas and carbon dioxide.

Municipal by-products: refer to biomass wastes produced by the urban population and consist of two types of products: solid municipal by-products and gas/liquid municipal by-products produced in cities and villages.

Solid municipal biofuels: comprise by-products produced by the residential, commercial, industrial, public and tertiary sectors that are collected by local authorities for disposal in a central location, where they are generally incinerated (combusted directly) to produce heat and/or power. Hospital waste is also included in this category.

Gas/liquid municipal biofuels correspond to biofuels derived principally from the anaerobic fermentation (biogas) of solid and liquid municipal wastes which may be landfill gas or sewage sludge gas.

3.4 Some Key Facts of Agriculture and Energy (Food and Agriculture Organization (FAO) 2000a)

– Globally, the agriculture production (food, forage, fiber, forest and pharma, medicine decoration plants) consumes 30% of the world's available energy—with more than 70% consumed beyond the farm gate.
– The agriculture production produces about 31% of the world's greenhouse gas emissions.
– Almost one-third of the food we produce is lost or wasted. This accompanying with another loss in energy reached about 38% of the energy consumed in the agro-production chain, as seen in Figs. 3.6 and 3.7.

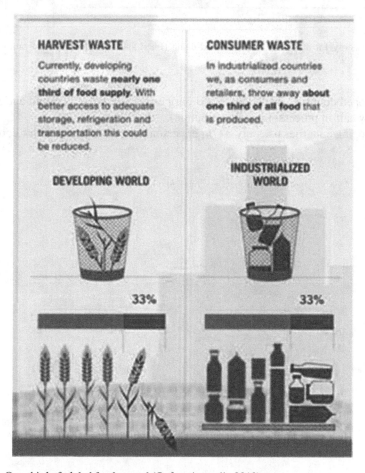

Fig. 3.6 One-third of global food wasted (Oxfam Australia 2012)

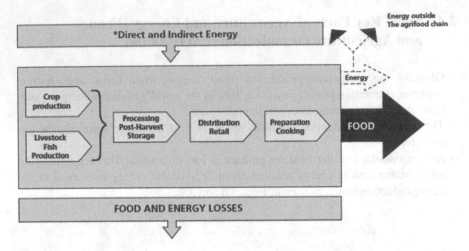

Fig. 3.7 Energy for and from the agro-food chain (Food and Agriculture Organization (FAO) 2000a)

- Developed countries use about 35 GJ per person a year for food and agriculture (nearly half in processing and distribution).
- Developing countries use only 8 GJ per person per year (nearly half for cooking).

Chapter 4
Energy in Agriculture

4.1 Integration Between Agriculture and Energy

In general, both agriculture and energy have double roles as user and supplier for each other. Agriculture supplies energy by wood, charcoal, vegetable and crop wastes as primary energy. Moreover, it supplies biogas and biofuel with both ethanol and diesel as clean and sustainable energy. On the other hand, there is a nexus between water, energy and food. Water and energy show that water supplies 15% of the global energy and consumes 17% of the energy supply. Agriculture residues also share in energy productions. Water, even fresh or saline, also shares in producing all types of natural gas, coal, crude oil, atomic energy, wind and solar and biofuels, as shown in Fig. 4.3.

Energy in agriculture has two roles, both direct and indirect. The direct use is as a fuel for tractors and all agriculture machines, irrigation pumps, drying crop equipment, cooling, processing, packaging, storage, transport, cooking and preservation of agriculture products, agro-industry and finally transportation of farmers to their field work as well as the transport of agriculture productions from fields to markets and factories. Energy also shares in wastewater treatments, water deliveries and water desalination and water and land reclamation and leveling. Energy is important as a main source of light for all agriculture product processes. The indirect uses of energy in agriculture are its roles of energy in agro-industry units such as producing chemical (such phytohormones and plant growth regulators as seen in Fig. 4.1), fertilizers and pesticides.

Energy is also required for increasing agricultural productivity, especially in developing countries; the energy needs to develop rural communities in CO_2 mitigation and adaptation by reducing carbon emission from biomass and from using biofuel instead of fossil fuels (Figs. 4.2 and 4.3).

© The Author(s), under exclusive license to Springer Nature Switzerland AG 2020
N. Noureldeen Mohamed, *Energy in Agriculture Under Climate Change*,
SpringerBriefs in Climate Studies,
https://doi.org/10.1007/978-3-030-38010-6_4

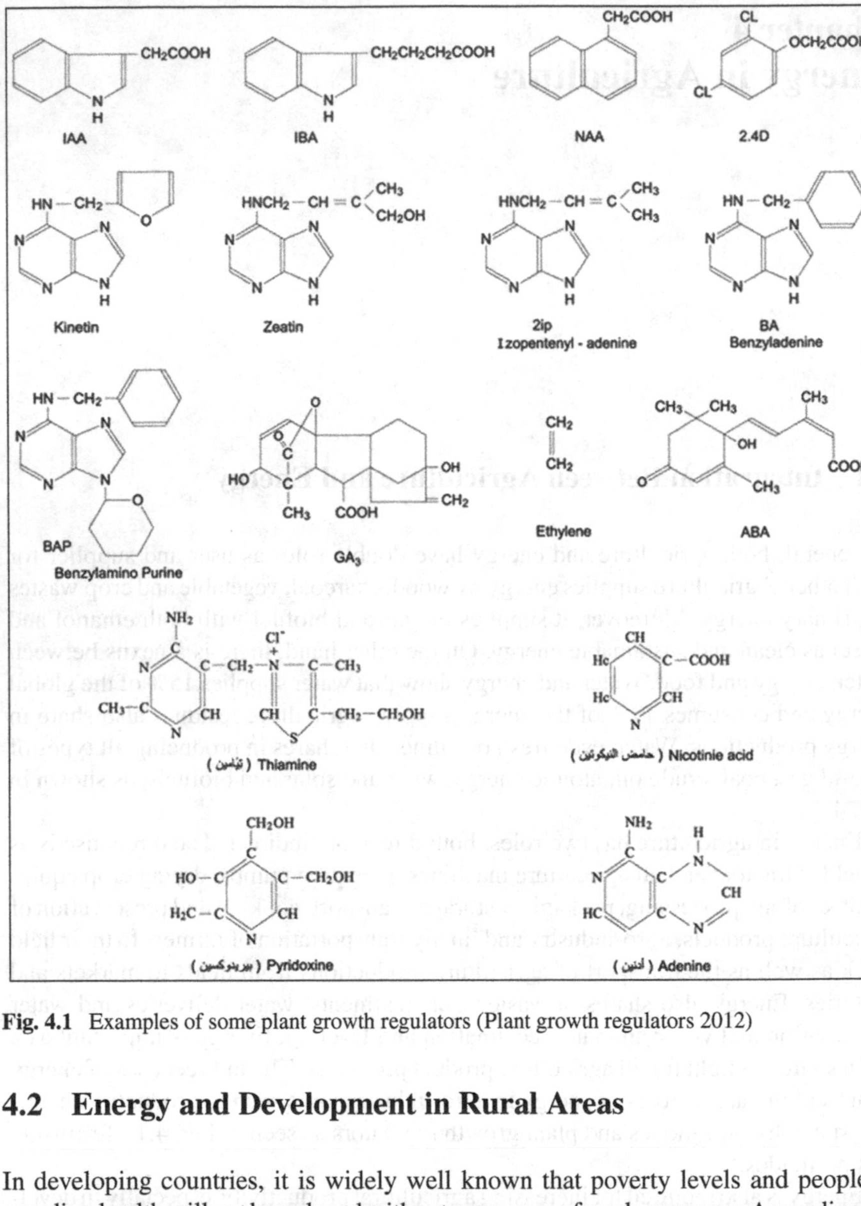

Fig. 4.1 Examples of some plant growth regulators (Plant growth regulators 2012)

4.2   Energy and Development in Rural Areas

In developing countries, it is widely well known that poverty levels and people's receding health will not be reduced without more uses of modern energy. According to WB 2000 (Bundschuh, University of Southern Queensland (USQ), 2014) surpassing the one toe/capita per year level of energy use seems to be an important instrument for development and social change. While low energy consumption is not the only cause of poverty and under-development, it does appear to be a close proxy for many of its causes. For example, environmental degradation, poor health care, inadequate water supplies and female and child hardship are often related to low energy consumption.

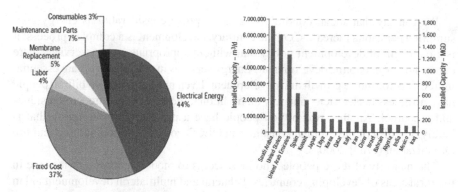

Fig. 4.2 Energy uses in different sectors related to agriculture (Cooley et al. 2006; Food and Agriculture Organization (FAO) 2000)

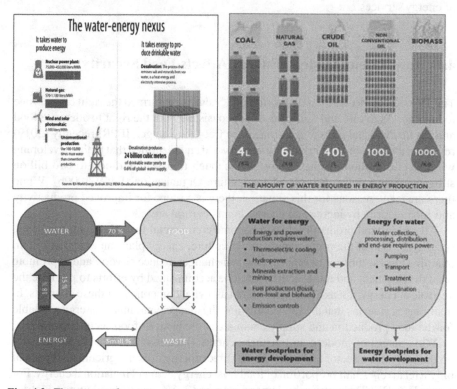

Fig. 4.3 The nexus of water, energy and food (Cooley et al. 2006; Food and Agriculture Organization (FAO) 2000)

We can say that social conditions surely improve considerably as energy consumption per capita increases. On the contrary, development is a complex process; it is a paradigm of development policy that without appropriate energy services there can be no true economic development. Energy services in suitable forms are essential ingredients for future growth and development. Even now, around two billion people have no access to electricity and rely on traditional fuels, such as dung, crop residues and woodfuel. Another two billion people have a per capita consumption that is barely one-fifth of the average consumer in OECD countries (Food and Agriculture Organization (FAO) 2000a).

The majority of these people who lack access to modern energy services are in the rural areas of developing countries. Bilateral and multilateral development aid in support of national efforts has included a variety of rural energy programs, including investment projects, training and capacity building, trying to improve the provision of energy services.

4.3 Insufficient Modern Energy Affects Food Security

The 1996 Rome Declaration on World Food Security reaffirmed the right of everyone to have access to safe and nutritious food, consistent with the right to adequate food and the fundamental right of everyone to be free from hunger. IFPRI recently (2019) reviewed the state of food insecurity in the world and identified that in the developing world, 811.6 million people do not have enough to eat, in addition to two billion suffering from poverty (Food and Agriculture Organization (FAO) 2000b). While there are many reasons for this situation, there is a scope for improved productivity and food security by increasing energy inputs in rural areas.

All agricultural production has to become commercial to enable long-term food security; as one of the prime inputs to agriculture, energy plays an essential role in attaining food security. Attempts to eliminate hunger, reduce poverty and to promote rural development and food security must be accompanied by efforts to promote the key role of energy, not as a goal in itself but as a vital component of these attempts. In many cases, it seems that the low quality and the meager amounts of energy available for the food production and supply chain are at the heart of food security problems.

Wide ranges of modern and primary energy (biomass) forms are used directly on the agriculture sector and rural areas. Direct energy uses in agriculture represent only a relatively small proportion of total final energy demand in national energy. For example, in OECD countries the demand is around 3–5%, and in developing countries between 4 and 8% (Bundschuh, University of Southern Queensland (USQ), 2014). Energy for agricultural practices in many developing countries continues to be based to a large extent on human and animal energy, and on traditional woodfuels. Empirical evidence suggests that the potential gains in agricultural productivity through the deployment of modern energy services are not being fully realized in developing countries.

4.4 Food Price and Energy

Starting from 1960s and 1970s, the progress in industry solved the global food shortage problem at the time, not only through conventional plant breeding but also by tripling the application of inorganic fertilizers (chemical fertilizers), expanding the land area under irrigation, and increasing direct and indirect fossil fuel energy inputs to provide additional services along the agriculture food chain, including mechanization, agriculture chemicals and many others. Nowadays, the annual incremental yield increases of major cereal crops are in decline and some fossil fuels are becoming relatively scarce and hence costlier. The following figure shows the relation and correlation between food prices and crude oil prices as is the correlation of the broader energy commodity prices with food prices.

Partly due to fluctuating world energy prices reaching a peak, a related increase in global food prices was evident in 2008 which hit developing countries the hardest. In the poorest urban households where food can account for 60–80% of total expenditure, this increase resulted in societal unrest in some regions. In an average household in developing countries, the food bill is typically 7–15% of total expenditure, though food price rises are still of concern.

The global financial recession in 2008 had a significant impact on these commodity trends which reduced prices significantly before rising again.

Global climate change policy has been affected as a result of these concerns. Where climate policies have been connected to broader economic and energy security issues as in Australia, South Korea, Brazil and even in Europe, it has proved relatively easy to still achieve some political action for GHG mitigation. Conversely, where economic and security concerns have tended to dominate, as in USA and developing countries, climate change policies have been largely neglected (ADB 2009; GEA 2012; IEA 2007).

Oil price also affected all input of agriculture such as chemical fertilizers and pesticide, as shown in Figs. 4.4, 4.5, 4.6 and 4.7 (U.S. Department of Agriculture, Economic Research Service (ERS) 2011).

4.5 Energy in Irrigated Agriculture

Rain-fed agriculture is the main source of the world food production. Irrigation plays an important role as irrigated agriculture is used worldwide, especially in low rain and water shortage countries. Even irrigated agriculture does not exceed 20% of the total world cultivated lands but it produces 40% of the world's food and other agro-production. The share of global rain-fed agriculture is 60% of world food and agriculture production.

Under rain-fed conditions, crop growth is subject to the random variability of rainfall in space and time, or as we say risky conditions. The variability of rainfall over time increases with the decreasing annual and seasonal rainfall levels, especially

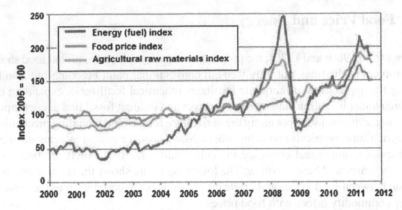

Fig. 4.4 Trends of commodity price indices for fuel, food and agriculture raw material from 2000 to 2012 (Mundi 2012)

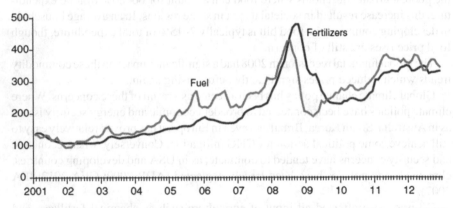

Fig. 4.5 Index of prices paid by agriculture producer in the USA from 2001 to 2012 (U.S. Department of Agriculture, Economic Research Service (ERS) 2011)

after climate change and global heating conditions. Rainfall variability is strongly related to crop yields in rain-fed tropical agriculture, particularly in semi-arid and dry sub-humid areas (rainfall between 400 and 900 mm/a) where water is a major constraint in food production (FAO (Food and Agriculture Organization of the United Nations) 2010, 2011).

The main disadvantages of rain-fed agriculture are (FAO (Food and Agriculture Organization of the United Nations) 2010, 2011, 2012):

- Floods and falling rains
- Vigorous weeds expansion
- High disease and pest prevalence (mostly according to high temperature with high humidity)
- High cost of farm inputs such as fertilizer and pesticides
- High post-harvest losses (mostly under primary cultivation)

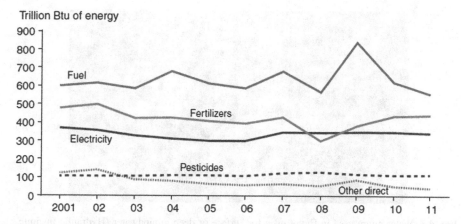

Fig. 4.6 Relation between fuel and electricity prices and chemical fertilizers and pesticides prices in the USA (Mundi 2012; U.S. Department of Agriculture, Economic Research Service (ERS) 2012)

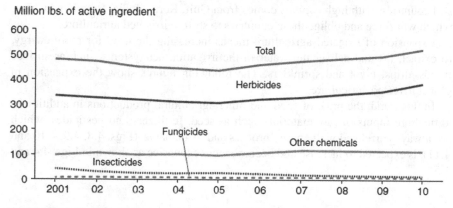

Fig. 4.7 Pesticide use in USA crops from 2001 to 2010 (U.S. Department of Agriculture, Economic Research Service (ERS) 2012)

- Weak extension services
- Lack of credit
- Inadequate information on market opportunities and a weak experience in the market for the farmer.

For the last disadvantage, the importance of irrigated agriculture is nominated to increase and will replace little by little the rain-fed agriculture due to their high and stable productivity while minimizing risky conditions. Moreover, the effects of global heating in shifting the rain areas to the north and the possibility of more frequent droughts, floods and other extreme events, especially under the prediction that estimates a reduction in the global water resources by 6% with less quality, will accelerate the transfer from rain fed to irrigated agriculture. Moreover, the foreign

Fig. 4.8 Water pump used in Egypt even for surface or deep groundwater (Hydraulic Institute, Europump, and The U.S. Department of Energy 2004, and by author)

investments in some developing countries, especially in Africa rich with water and soil countries, with high capitals comes from Gulf, Korea, India, Brazil and China, which will force and oblige these countries to shift to irrigated agriculture.

Expansion of irrigated agriculture means increasing the need for more energy, to extract, lift and deliver the water in the irrigation net systems, such as surface floods, drips, pivot and sprinklers. The following figures show the expansion in global irrigated agriculture.

In this trend, the price of food and other agricultural productions in addition to agriculture inputs of raw materials such as seed, fertilizers and pesticides, which are always correlated with energy process and its demands (Figs. 4.8, 4.9, 4.10 and 4.11), is expected to increase and become more expensive as shown in Figs. 4.4 and 4.5.

4.6   Energy in Water Resources

Water, not the soil, is the main governing key for agriculture activity. Irrigation has been a main driver which leads the green revolution and continues to play a key role in producing food and other agriculture productions (food, fiber, forage, forest and pharma, aromatic, medical and decorative plants) (Comprehensive Assessment 2007). Globally, some 300 million hectares of cultivated lands, representing 20% of the total world's cultivated land, are irrigated. Irrigated agriculture, which accounts for 70% of all freshwater withdrawals, contributes 40% of the world's food production (Hydraulic Institute, Europump, and The U.S. Department of Energy 2004). Inaccessible water supplies have a significant impact on the productivity of on-farm irrigation systems. A global estimate is that 40% or more of the water withdrawn

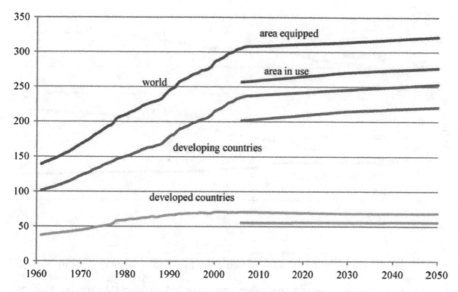

Fig. 4.9 Arable irrigated land: equipped and in use (million ha) (Mushtaq and Maraseni 2011)

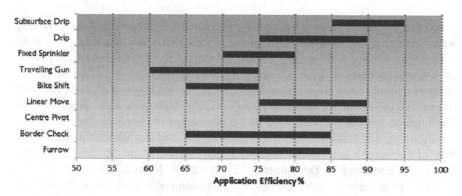

Fig. 4.10 Application efficiencies for various irrigation systems (Alexandratos and Bruinsma 2012)

for irrigation is "lost" through weak leakage, evaporation, deep percolation or surface run-off (Hydraulic Institute, Europump, and The U.S. Department of Energy 2004). Most of these "losses" are in fact recovered through tube-wells and shallow lift pumping from drainage channels or water table. Even if water is supplied effluent, distribution and delivery under gravity, the actual benefits of water distribution and storage through irrigation schemes involve an increase in energy intensity per hectare of cropped land. Water management in agriculture faces a number of challenges that will affect the availability and reliability of future water resources (Hydraulic Institute, Europump, and The U.S. Department of Energy 2004). Improving irrigation efficiency in agriculture is essential to save the precious and costly water (especially

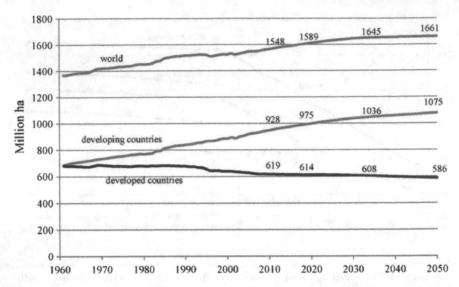

Fig. 4.11 Arable land and land under permanent crops: past and the future (Mushtaq and Maraseni 2011)

deep groundwater). Energy consumption in irrigated agriculture correlated primarily with water pumping requirements because it continuously works during all growing seasons (mostly for 4–6 months). Estimates suggest that by the end of the twentieth century, as much as 20% of energy worldwide was used by pumps of various types (Comprehensive Assessment 2007). This high-energy use explains the need to improve the efficiency and high performance of both pumps and pump motors. Even though there is no breakdown of the percentage of energy that used to pump groundwater, it is estimated that it may be 1 or 2%, and almost certain that at least 75% of this usage was for pumping groundwater for agriculture (Hydraulic Institute, Europump, and The U.S. Department of Energy 2004; Jones 2012).

As the population growth increases, water consumption also increases to produce more foods and other services and products needed to provide the essential five food groups which include meat, sugar, grain and cereal, oil and milk. These five groups include 55 food items that are consumed daily by most of the people in addition to another three folds of this number as supplementary food and goods. For example, the cereal group includes rice, wheat, barley, oats, coarse and fine grain such as maize. The same is with other main essential food groups also. Figure 4.12 and Table 4.1 show the increase in global water consumption from year 1900 until year 2025 (projected).

The relation between the energy elements and water operation, extraction and uses are shown in Table 4.1.

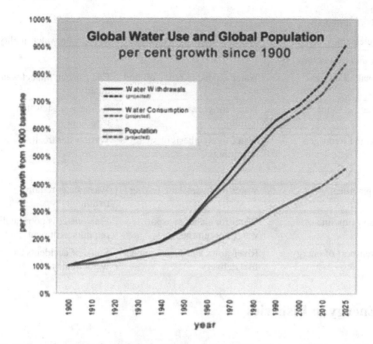

Fig. 4.12 Increase in global water consumption from year 1900 until year 2025 (projected) (United Nations Environment Programme (UNEP) 2012)

Table 4.1 Energy and water in agriculture (Murray 2008)

Energy element	Implications of water quality	Impacts on water quality
Energy extraction and production		
Oil and gas exploration	Water for drilling, completion and fracturing	Impact on shallow groundwater quality
Oil and gas production	Large volume of produced, impaired water	Produced water can impact surface and groundwater
Coal and uranium mining	Mining operations can generate large quantities of water	Tailing and drainage can impact surface water and groundwater
Electric power generation		
Thermoelectric (fossil, biomass, nuclear, solar thermal)	Surface and groundwater for cooling and scrubbing	Thermal and air emissions from impact surface waters and ecology
Hydroelectric	Reservoirs loose large quantities to evaporation	Can impact water temperatures, quality, ecology
Solar PV and wind	None during operation, mining water use of panel and blade washing	

(continued)

Table 4.1 (continued)

Energy element	Implications of water quality	Impacts on water quality
Refining and processing		
Traditional oil and gas refining	Water needed to refine oil and gas	End use can impact water quality
Biofuel and ethanol	Water for growing and refining	Refinery wastewater treatment
Synfuels and hydrogen	Water for synthesis or steam reforming	Wastewater treatment
Energy transportation		
Energy pipelines	Water for hydrostatic testing	Wastewater requires treatment
Coal slurry pipeline	Water for slurry transport, water not returned	Final water is poor quality, requires treatment
Barge transport of energy	River flows and stages impact fuel delivery	Spills of accidents can impact water quality

4.7   Energy in Fisheries

Globally, fish represents 16% of animal protein, and 6.1% of all proteins consumed by people. Annual per capita consumption has grown from 11.5 kg in the 1960s to 17.1 kg in 2008 (Jones 2012). Capture fishing has traditionally been the primary source of supply, but it is facing resource and yield limits. Aquaculture has significantly increased over the last two decades. Nowadays, it contributes to some 60–70% of food fish supply (FAO 2011).

Fish consumption varies across regions and countries. This variety reflects the different levels of availability of fish and other foods, as well as diverse food traditions, demand, income levels, lifestyle and prices. Developed countries have generally higher consumption levels and have increasingly looked on imports to satisfy the demands. In developing countries, changes in consumption have been more variable. For example, in sub-Saharan Africa, consumption has been relatively low and, in some cases, it is declining. However, broad projections suggest that global per capita consumption will rise (Food and Agriculture Organization of the United Nations (FAO) 2012).

In both capture fisheries and aquaculture, a wide variety of technologies, from artisanal to highly industrial, are involved in both production and supply. These different technologies, which encompass vessels, equipment and culture systems, use a range of different types of energy in varying amounts either (Food and Agriculture Organization of the United Nations (FAO) 2012). Because of increased modern technology fishing vessels and increased numbers of fishers, the intensification of aquaculture, and growth in processing, transport and retail market distribution, fossil fuels have played an increasingly important role in fisheries production and distribution. The economic viability of the sector has become closely linked to fuel and energy prices and their indirect impacts on key inputs, such as aquaculture fertilizers and fish feeds (Food and Agriculture Organization of the United Nations (FAO) 2012). Most of the current production methods originated when resources were abundant, energy costs were dramatically lower and less attention was paid to operating efficiency and ecosystem impacts. The new realities of high energy prices and greater environmental awareness present major challenges for the future viability of the sector. This may be especially true in developing countries where access to and promotion of energy-efficient technologies has been limited (FAO 2011; Suuronen et al. 2012). Future production will be increasingly constrained by the cost of fuel and energy supplies. The expansion of aquaculture will need to make efficient use of embodied energy in feeds and to use energy efficiently in production systems. The fishing industry will need to become energy-smart along the entire food chain to cope with the volatility and rising trends of fuel and energy prices (Food and Agriculture Organization of the United Nations (FAO) 2012) and to ensure food availability at accessible prices.

4.8   Energy in Aquaculture

The use of fuel and energy in aquaculture is more indirect than in capture fisheries. Aquaculture production systems are diverse, ranging from low-intensity subsistence operations to high-intensity industrial models. The rapid growth of aquaculture production has been accomplished in part through intensification. Global feed production for culture fish and crustaceans is estimated at about 6 million tons (FAO 2007).

Along with the energy costs of capturing the feedstock, fish production meal and fish oil require significantly high energy for cooking and drying. Energy consumption in fish meal production is estimated at 32 kilowatt-hours (kWh) plus 32 l of fuel oil per ton of other raw materials processed (FAO 2007). Substantial parts of the aquaculture sector are now reducing their dependence on fishmeal and oil, but the energy required to produce terrestrially sourced raw materials is still significantly high. Water exchange, replacements and treatment, boats, vehicles and handling systems need appreciable amounts of energy demands in aquaculture. However, these demands are usually less important than feeds and fertilizers. Growth in the aquaculture sector will depend on improving feeding efficiency and increasing land or water-based unit productivity. Information on energy options and strategies for feeding and practical advice for producers will be important in supporting the aquaculture sector and ensuring that it provides sustainable benefits for producers and consumers.

4.9   Energy in Food Industry

Nowadays, 95% of food production comes from agriculture, and only 5% may come from hunting and open fishing. Thus, I'm not satisfied or convinced with the new term used in Canada and some other countries about the **"agro-food"** with meaning of "**the food that is produced from the agriculture sector**". These terms mean we may also use agro-fiber, agro-forage, agro-forest and agro-pharma, agro-aromatic and so on!! Thus we can almost say that each food has an agriculture origin, even the open fishing and hunting, because it is studied in most faculties and schools of agriculture and farmers around the world. The same trends in hunting and the way of strips, scalp scaling, fly and cooking are taught in faculties and school of agriculture.

Agro-food and agro-industry are in progress, and the selling of most supermarkets according to World Bank (2010), Verma (2008) reached 65–80% from the total sells. This includes different manufactured food of all types, such as packed, cooling, freezing, semi- or full-cooked, fried, grilled, dairy products, cereal, legumes, oils and so on. This growth is due to increase of educated people, especially in the developing countries, and the needs of working of both wife, husband and most of family members to follow the soaring of prices of all products in addition to the high poverty level in these communities. Growth of agro-industry means the growth of energy needed and energy consumed especially under high temperatures which will increase as a result of global heating and the fast spoiling and damage of most vegetable, fruits, milks and dairy products.

4.10   Energy Consumption in Agro-Industry

Energy with its double faces, electricity and fuels (liquid and gas) are used in the following operation during food industry:

Electricity: lighting—water heater—milking plant—motor pump—refrigerator—waste heat—milk cooking—effluent disposal—irrigation—fertilizers and pesticides manufactures, and so on.

Liquid and gas fuels: tractors—fertilizer trucks—topdressing planes—cultivation and establishment—spraying—forage conservation—track and road repairs—pasture renovation—drainage (both electricity and fuel)—feeding out—transportation—other (CAE 1996). The following are some of the examples of energy consumption in industrial processes:

- Concentrate dilutes liquid foods such as milk and juice;
- Remove microbes in liquid foods for cold pasteurization;
- Recover dissolved solids such as sugars from processing wastewater; and
- Treat processing wastewater.

The four key elements relating to energy-smart food, forage, fiber and agro-pharma (medical, aromatic and other plants) productions, sustainable agriculture and rural development are:

- having the right energy mixes to locates the ambition to produce 60% more food and pre-food (such as dry forage to feed livestock's and then to feed people) by 2050;

- consolidating rural and village livelihoods in a sustainable way by provision of basic energy services;
- providing non-expenses and secure energy sources (especially through renewable energy systems when possible) now and in the future; and
- addressing how the whole agro-food industry could become more sustainable and energy-smart and efficient.

This implies using low-carbon emission energy systems in high efficient manner, providing greater energy access, and enabling the achievement of national food production and sustainable development goals.

Figures 4.13, 4.14, 4.15, 4.16, 4.17 and 4.18 show some examples of using energy in different agro-industry activity:

Fig. 4.13 Breakdown of typical electricity use in a typical non-irrigated dairy farm (CAE 1996)

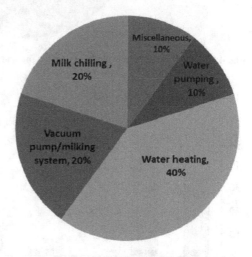

Fig. 4.14 Liquid fuel demands for a typical sheep and beef farm (CAE 1996)

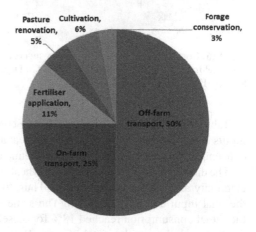

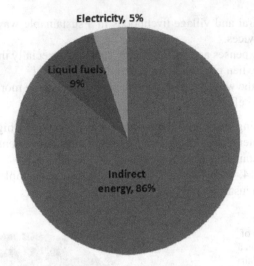

Fig. 4.15 Energy inputs into typical intensive pig and poultry production enterprises (CAE 1996)

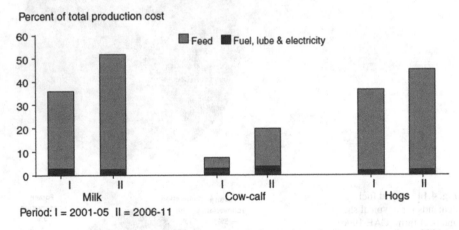

Fig. 4.16 Cost shares feed and direct energy inputs in livestock production expenses from 2001 to 2005 and from 2006 to 2011 (Mundi 2012; U.S. Department of Agriculture, Economic Research Service (ERS) 2012)

Tables 4.2, 4.3 and 4.4 show the energy consumption in the USA as agricultural inputs in some essential crops, such as wheat and soybean, in addition to the energy consumption uses in producing major chemical fertilizers needed for crops (NPK).

The diesel consumption as input in wheat production represents 24% + 10% for electricity and 2% for transportation. Thus, the energy represents a total of 36% of the total input cost of cultivation. The same rates are observed in soybean where the diesel consumption reached 18% for diesel +1% for electricity and another 1% for transportation, with a total of 20%. Both examples are for rain-fed agriculture

Fig. 4.17 Breakdown of
direct and indirect energy
inputs for a typical New
Zealand apple orchard (CAE
1996)

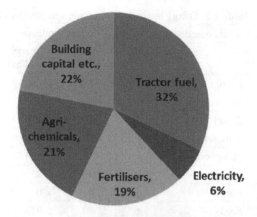

Fig. 4.18 Main liquid fuel
end uses on a typical arable
cropping farm (CAE 1996)

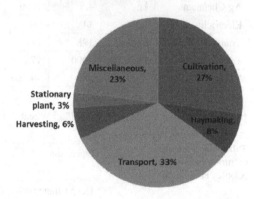

without cost of irrigation, and without adding the cost energy used in producing
fertilizers, pesticides transportation and marketing.

Regarding the energy consumption to produce the major nutrients as chemical
fertilizers of nitrogen, phosphate and potassium reflects that the nitrogen production
is heavier energy consumer than phosphate and potassium and is represented eight-
fold (time). Moreover, there is energy consumption in producing pesticides hormones
and growth regulators. Figures 4.19, 4.20 and 4.21 show the correlation between
energy and agriculture in producing fertilizers and pesticide in addition to increase
demands on electricity periodically and the energy to and from the food value chain.

Figure 4.22 shows the energy to and from food value chain.

Table 4.2 Energy inputs to US farming systems (IEA 2011; Pimentel and Pimentel 2009)

Inputs quantity		Wheat production			Soybean production		
	Unit	Qty (ha^{-1})	Energy (MJ/ha)	(%)	Qty (ha^{-1})	Energy (MJ/ha)	(%)
Labor	h	7.8	1005	6	6	1005	9
Machinery	kg	50	387.3	23	20	1507	14
Diesel	L	100	4187	24	33.8	1859	18
Nitrogen	kg	68.4	4580	27	3.7	247	2
Phosphorus	kg	33.7	599	3	37.8	653	6
Potassium	kg	2.1	25	0	18.8	201	2
Lime	kg	0	0	0	2000	2352	22
Seeds	kg	60	913	5	56	1884	18
Ag Chemicals	kg	4.5	1696	10	1.7	712	7
Electricity	KWh	14	50	0	10	121	1
Transport	mg km	198	281	2	150	67	1
Total	kg	2900	17,208	100	2600	10,608	100
Yield			44,660			38,740	
MJ output/input			2.6			3.65	

embodied energy Machinery 77.5 MJ/kg; Wheat = 15.4 MJ/kg; Soybean = 14.9 MJ/kg

Table 4.3 Fuel use in common crop operation (Cooley et al. 2006)

Operation	Fuel use (L/ha)	
	Range	Mean
Deep tillage (subsoiling)	15–25	20
Plowing (moldboard of disc)	15–20	18
Chisel or sweep tillage	7–12	9
Shallow tillage or seeding	3–10	5
Spaying	0.5–2	1.5
Grain harvesting (combine)	8–15	10
Forage harvesting (chopper)	10–20	15
Cotton harvesting (picker)	12–18	15

Table 4.4 Energy embodied in fertilizers (Pimentel and Pimentel 2009)

Fertilizers	MJ/kg
Nitrogen	75.63
Phosphate (P_2O_5)	9.53
Potassium (K_2O)	9.85
Sulfur	1.12

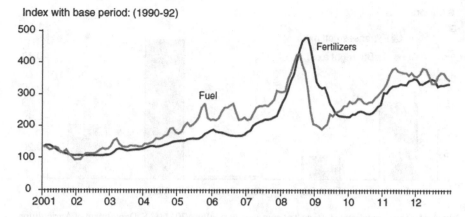

Fig. 4.19 Prices paid by agriculture producers from 2001 to 2012 (Mundi 2012; U.S. Department of Agriculture, Economic Research Service (ERS) 2012; FAO/USAID 2015)

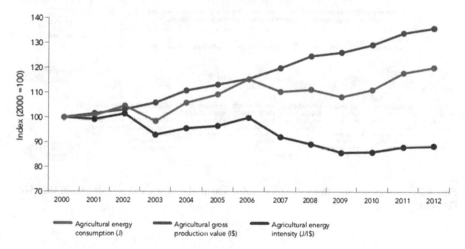

Fig. 4.20 Agriculture is continuing to demand higher level of energy (Zentner et al. 2004; FAO/USAID 2015)

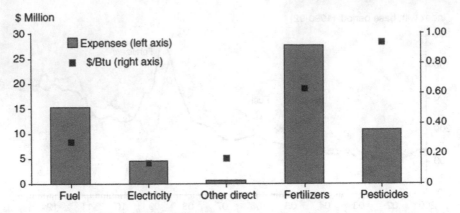

Fig. 4.21   The expanding in total cost of energy in agriculture 2011 (U.S. Department of Agriculture, Economic Research Service (ERS) 2012)

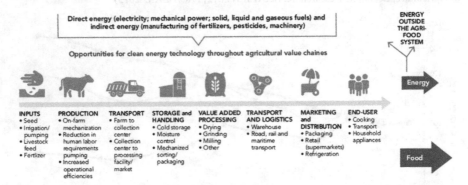

Fig. 4.22   Energy to and from the food value chain (http://www.fao.org/docrep/014/i2454e/i2454e00.pdf)

Chapter 5
Water Energy Food Nexus

5.1 One Nexus

The correlation and integration between water, energy and food (WEF) comes true and becomes so close. We will never use the terms again, such as "the future of food production", or the "future of energy security", or the future of water demands. These three items have become one nexus that includes and covers water, energy and food. Moreover, with the climate change impacts that already exist, the nexus accepts them and consists of a fourth item and soon will also accept the population growth to become a pentagonal nexus. Thus starting from the year 2018 the world started to use the term "future of nexus of water, energy, food, climate change and population growth". This is only logical because no agriculture production exists without energy, no food without water and no energy without water and food, and the global population will be 9.6 billion capita in year 2050 instead of 7.6 billion in 2018. Thus we are talking about one integrated nexus with no possibility to separate any of its items.

   With global heating, FAO estimated that the food productions should be increased by 50% before 2050 from only 20–30% more water. This means the amount of agricultural chemicals such as fertilizers, pesticides, growth regulators and hormones will increase to meet the need of the increasing demand of food and other agriculture productions by 60%. The climate condition will cause a decrease of available water by 6%, as well as lowering its quality may increase the energy consumption to double before the year 2100 and by 50% by year 2030, as shown in Figs. 5.1, 5.2, 5.3 and 5.4. Thus the traditional nexus of water, energy and food is modified to include climate change (water, food, energy and climate change) and I suggested in the last FAO meeting in June 2018 for Soil Partnership Plenary assemble that it can be modified again to include population growth (water, food, energy, climate change and population growth), in addition to call of: more food from less water. This means

© The Author(s), under exclusive license to Springer Nature Switzerland AG 2020
N. Noureldeen Mohamed, *Energy in Agriculture Under Climate Change*,
SpringerBriefs in Climate Studies,
https://doi.org/10.1007/978-3-030-38010-6_5

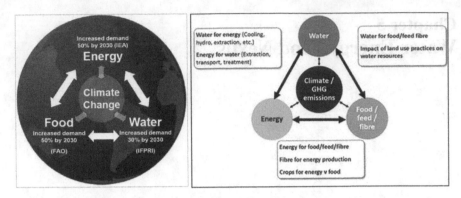

Fig. 5.1 Nexus of water, energy, food and climate change up to 2050 (Food and Agriculture Organization of the United Nations Rome 2014; Hoff 2011; World Economic Forum WEF. 2011)

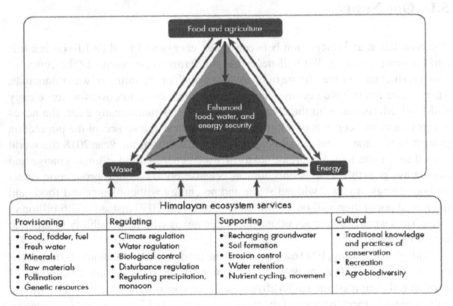

Fig. 5.2 Nexus of energy, water, food and climate change framework development (IISD REPORT FEBRUARY© 2 2001313)

urgent funding needs to agricultural research to produce high yield crops seeds and trees, and at the same time, consume less water and be heat-tolerant, thirsty, water shortage and dryness.

Figures 5.1, 5.2, 5.3, 5.4 and 5.5 show the last nexus of energy water food and climate change in the near future.

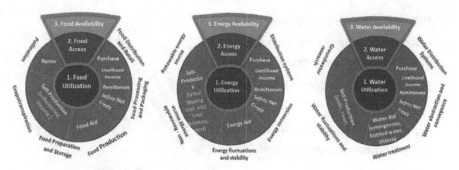

Fig. 5.3 Defining securities as core elements of water, energy and food nexus (IISD REPORT FEBRUARY© 2 2001313)

Present	**Future**
Unmet demands!	Increasing demands!
• 0.9 billion people have no access to clean water	• Due to population growth, economic development and changing consumption patterns
• 1.1 billion people have no access to electricity	• The uncertainties of global change exacerbate the difficulty in achieving these goals
• 1 billion people have insufficient food supply	

by **2035**
GLOBAL ENERGY
consumption will
INCREASE 50%

... increasing
WATER
CONSUMPTION
by **85%**

To meet global food demand by 2050, agricultural production must **INCREASE BY 60%**

NOW

2050

Fig. 5.4 Achieving water, energy, food security in a dynamic world (CGIAR n.d.; United Nations Department of Economic and Social Affairs (UNDESA) 2014)

5.2 Energy and Food

Figure 5.6 shows the relation between energy and food production. Food production and supply chains make approximately 30% of total global energy demands. On the other hand, the use of land productions from foods and wasted residues to produce biofuel share in producing appreciable amounts of energy.

Examples of interconnections:

- Agriculture accounts for 70% of total global freshwater withdrawals
- 90% of energy produced today is water-intensive
- Agriculture & food chain account for 33% of global energy demand

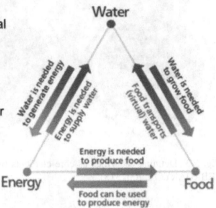

Fig. 5.5 Interconnection between WEF nexus (United Nations University (UNU) 2013; Nexus Regional Dialogue Programme (NRD) 2018)

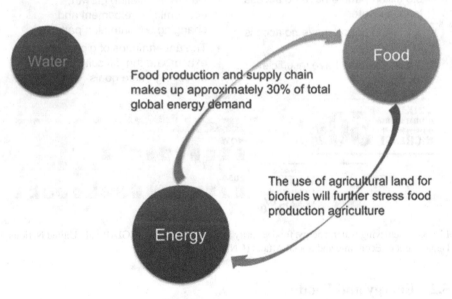

Fig. 5.6 Interconnection between energy and food (Nexus Regional Dialogue Programme (NRD) 2018)

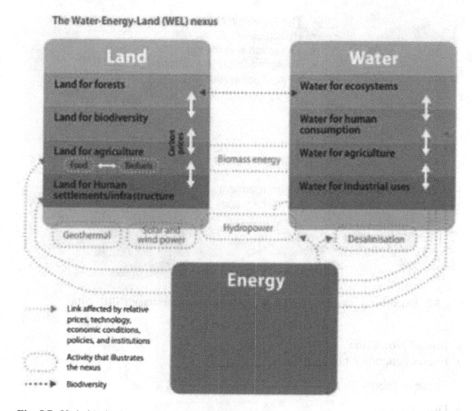

Fig. 5.7 Variation in the concept to use land instead of food (European Report on Development 2012)

5.3 Energy and Land Instead of Food

Because the soil produces more than 95% of the global food and agriculture production, we can say no food without soils (so far, under aquatic and hydroponic agriculture). The European report in development 2012 suggested the possibility of using "land" instead of "food" to get the same logic meaning as shown in Fig. 5.7.

5.4 Energy and Water

The relation between water and energy may be concluded as follows:
Energy in water:

- Thermoelectric cooling
- Hydropower
- Extraction and mining

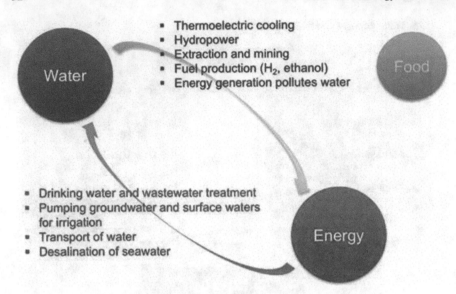

• Thermoelectric cooling
• Hydropower
• Extraction and mining
• Fuel production (H₂, ethanol)
• Energy generation pollutes water

• Drinking water and wastewater treatment
• Pumping groundwater and surface waters
  for irrigation
• Transport of water
• Desalination of seawater

Fig. 5.8 Water energy interaction (Nexus Regional Dialogue Programme (NRD) 2018)

• Biofuel production
• Energy generation polluted water.

  Water in energy:

• Drinking water
• Pumping ground and surface water for irrigation
• Transport of water
• Desalination of seawater
• Water treatments for drainage, industrial and sanitation water.

  Figure 5.8 shows the relation between water and energy.

5.5   Conservation Agriculture to Save Energy

This is an approach to manage ecosystems for improved and sustained productivity by minimizing mechanical soil disturbance, providing permanent soil cover to maintain moisture content and diversifying crop species grown in rotation. Reduced energy can result from less fuel used for tillage, less power for irrigation and less indirect energy needed for weed control per unit of produce.

5.5.1 Irrigation Water

Water pumping for drinking water, irrigation and food processing consumes a lot of energy, usually by the use of either electricity or diesel for internal combustion engines, to power the pumps. Solar and wind-powered pumps even if they are still more expensive than the traditional energy (Fig. 5.9) are growing in popularity and should be encouraged where the potential for solar and wind energy exists. Energy demands for irrigation can be reduced by:

- Using gravity supply as much as possible.
- Using high-efficient electric motors.
- Sizing pumping systems to meet crop's actual water requirements without loss.
- Choosing high-efficient water pumps that are correctly matched to suit the task.
- Giving a high priority to pump maintenance regularly.
- Using low-head distribution sprinkler systems or drip irrigation in row crops.
- Reducing water leakages in all components of irrigation systems.
- Monitoring soil moisture contents as a guide for water application.
- Choosing appropriate and drought-tolerant crop varieties.
- Using weather forecasts, especially for warm wind and high speed wind, to estimate the timing and needed water for the fields.
- Using global positioning systems (GPS) to varying irrigation rates across a field to match the soil and moisture conditions.
- Conserving soil moisture through covering the soil surface by mulch and tree shelter belts.
- Maintaining all equipment, water sources and intake screens in good working order.

Fig. 5.9 Field of solar panels in rural USA (Scientific American and E News 2019)

Fig. 5.10 Wind energy in agriculture (ET Energy World 2019)

5.5.2   Storage and Refrigeration

Cooling and cold storage are used to maintain food quality after harvesting, processing and to reduce losses along the supply chain. Refrigeration systems depend on reliable electricity supply systems, although new technologies such as solar absorption chillers are reaching the market. Other sources of renewable electricity can be used on both small and large scales. For cold stores, reducing energy demand is possible through such measures as increasing the insulation, keeping access doors closed and minimizing the heat load at the end of the processing phase of the cold chain (Figs. 5.9 and 5.10).

5.5.3   Fertilizers and Plant Nutrients Behavior

To understand the role of energy in manufacturing fertilizers as an essential element to agriculture, first we should be aware by how the plant takes the nutrients from the soil and in which forms, in addition to knowing how many nutrients are essential and needed to the plant life cycle and in which charge (positive or negative) will be available for the plants adsorption. The most important points in this discussion will be about the nitrogen element which is considered the most major element that the plant adsorbed from the soil to build the plant's body and is responsible for its green color. The critical issue in nitrogen demands needs to be explained, like the

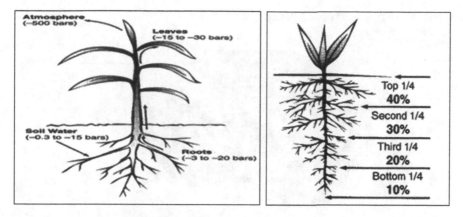

Fig. 5.11 Adsorption of nutrients from the soil solution by root zone (Library of Congress Cataloging In Publication Data: Foth 1990)

facts of the absence of any minerals or rocks in the soils containing nitrogen in their chemical formula can be weathered or decomposed to produce and release nitrogen. Thus, nitrogen should always be added to the soil from outer sources such as organic matter in the past before the industry revolution (or in organic farm recently) or as chemical nitrogen fertilizers. The other fact is that nitrogen element can be lost easily from the soil in the form of nitrogen gas (NO_2, NO, NH_3, etc.) or leached from the soil solution by deep percolation or seepage by the stream of rain or irrigation water as soluble nitrate (NO_3^-). This explains that there is no way to not add the nitrogen fertilizers to the soil frequently from outer sources in right recommended doses. The only way to reduce these amounts is to minimize the rate of nitrogen loss from the soil under different soil types, soil condition or warm and rainy weather regions. The next discussion will explain the behavior of the plant nutrition in the soils and the "plant soil water relationship" to reach optimum methods to deal with chemical fertilizers and reduce the amounts, as much as possible, that should be added to the soil to avoid the contamination of both soil and groundwater by the elements of chemical fertilizers, especially the nitrogen and trace elements.

Figure 5.11 shows, over the course of a growing season, that plants will extract about 40% of their water from the top quarter, 30% from the second quarter, 20% from the third quarter and 10% from the bottom quarter of the root zone.

5.5.4 Soil–Water–Plant Relationships

Most of the soil particles are charged mostly by negative charges and little site by positive charges, as shown in Fig. 5.12. This means the nutrient with positive charges will hold or absorb or get attracted to the negative charge sites of the soil particle, by attraction between positive and negative, to be as exchangeable cations holds by little

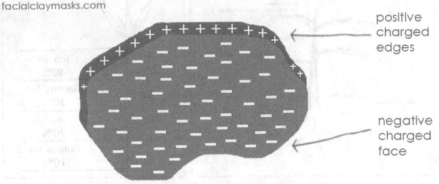

Fig. 5.12 Negative and positive charge amounts and sites in the fine soil particles (Library of Congress Cataloging In Publication Data: Foth 1990)

force power to be ready to be taken by the roots systems at any time (Figs. 5.12 and 5.13). This also means it will stay in the soil against gravity seepage (percolation) or lost by stream of rain or irrigation; vice versa, the negative charges nutrient will repel the negative charge of the soil particles and become so easy to move down with the water stream to be lost by deep seepage. The nitrogen form ready to be absorbed by the plant is the mono negative charge form as nitrate (NO_3^-) to become easy to be leached from the root zones. In some crops that are cultivated under water ponding such as rice, ammonium form NH_4OH may also be absorbed by the root systems, and also may be lost in the form of NH_3 gas or NO and N_2O, especially in the warm weather or in the sandy and calcareous soils.

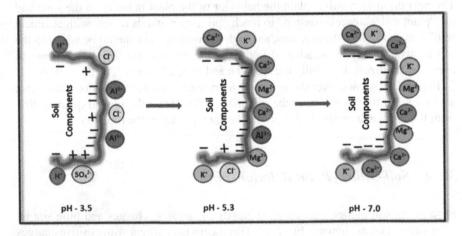

Fig. 5.13 Site of cation absorbs as cations exchangeable in the fine soil particles and colloidal with different soil pH values (Library of Congress Cataloging In Publication Data: Foth 1990)

Plants need certain essential nutrient elements to complete their life cycle. No other element can completely substitute for these elements. At least 16 elements are currently considered essential for the growth of most vascular plants. Carbon, hydrogen and oxygen are combined in photosynthetic reactions and are obtained from **air and water**. These three elements compose 90% or more of the dry matter of plants. **The remaining 13 elements are obtained largely from the soil** (Library of Congress Cataloging In Publication Data: Foth 1990).

Nitrogen (N), phosphorus (P), potassium (K), calcium (Ca), magnesium (Mg) and sulfur (S) are required in relatively large amounts and are referred to as the **macronutrients**. Elements required in considerably smaller amount are called the micronutrients. They include boron (B), chlorine (Cl), copper (Cu), iron (Fe), manganese (Mn), molybdenum (Mo) and zinc (Zn). Cobalt (Co) is a **micronutrient** that is needed by only some plants. Plants deficient in an essential element tend to exhibit symptoms that are unique for that element. More than 40 other elements have been found in plants. Some plants accumulate elements that are not essential but increase growth or quality. The absorption of sodium (Na) by celery is an example, and results in an improvement of flavor. Sodium can also be a substitute for part of the potassium requirement of some plants, if potassium is in low supply. Silicon (Si) uptake may increase stem strength, disease resistance and growth in grasses. Most of the nutrients in soils exist in minerals and organic matter. Minerals are inorganic substances occurring naturally in the earth. They have a consistent and distinctive set of physical properties and a chemical composition that can be expressed by a formula. Quartz, a mineral composed of SiO_2, is the principal constituent of ordinary sand. Calcite ($CaCO_3$) is the primary mineral in limestone and chalk and is abundant in many soils. Orthoclase-feldspar ($KAlSi_3O_8$) is a very common soil mineral, which contains potassium. Many other minerals exist in soils because soils are derived from rocks or materials containing a wide variety of minerals. Weathering of minerals brings about their decomposition and the production of ions that are released into the soil water. Since silicon is not an essential element, the weathering of quartz does not supply an essential nutrient, and plants do not depend on these minerals for their oxygen. The weathering of calcite supplies calcium, as Ca^+, and the weathering of orthoclase releases potassium as K^+. The organic matter in soils consists of the recent remains of plants, microbes and animals and the resistant organic compounds resulting from the rotting or decomposition processes. Decomposition of soil organic matter releases essential nutrient ions into the soil water where the ions are available for another cycle of plant growth. Available elements or nutrients are those nutrient ions or compounds that plants and microorganisms can absorb and utilize in their growth. Nutrients are generally absorbed by roots as cations and anions from the water in soils, or the soil solution. The ions are electrically charged. Cations are positively charged ions, such as Ca and K^+, and anions are negatively charged ions, such as NO_3^- (nitrate) and H_2PO4^- (phosphate). The amount of cations absorbed by a plant is about chemically equal to the amount of anions absorbed. Excess uptake of cations, however, results in excretion of H^+ and excess uptake of anions results in excretion of OH^- or HCO_3^- to maintain electrical neutrality in roots and soil. The

Table 5.1 Chemical formula
and common forms of the
essential elements absorbed
by plant roots from soils
(Library of Congress
Cataloging In Publication
Data: Foth 1990)

Nutrient	Chemical Symbol	Forms commonly absorbed by plants
Macronutrients		
Nitrogen	N	$NO_3^- - NH_4^+$
Phosphorus	P	$H_2PO_4^- - HPO_4^{2-}$
Potassium	K	K^+
Calcium	Ca	Ca^{++}
Magnesium	Mg	Mg^{++}
Sulfur	S	SO_4^{2-}
Micronutrients		
Manganese	Mn	Mn^{++}
Iron	Fe	Fe^{++}
Boron	B	$H_3BO_3^-$
Zinc	Zn	Zn^{++}
Copper	Cu	Cu^{++}
Molybdenum	Mo	MoO_4^{2-}
Chloride	Cl	Cl^-

essential elements that are commonly absorbed from soils by roots, together with
their chemical symbols and the uptake forms, are listed in Table 5.1.

In nature, plants accommodate themselves to the supply of available nutrients.
Seldom or rarely is a soil capable of supplying enough of the essential nutrients to
produce high crop yields for any reasonable period of time after natural or virgin lands
are converted to cropland. Thus, the use of animal manures and other amendments
to increase soil fertility (increase the amount of nutrient ions) is an ancient soil
management practice.

Energy embedded in the production of inorganic fertilizer (including nitrogen,
phosphorous and potash as NPK blends) is significant in addition to the chelate or
minerals micronutrients.

5.5.5 Adding Fertilizers in Sustainable Way

Farmers can save indirect energy by reducing the amount of fertilizers applied and
more accurate application methods as the following recommendations:

- Growing nitrogen-fixing legume crops as green crops in rotation with other crops
  to avoid nutrient depletion.
- Selecting an NPK fertilizer according to leaves and soil analysis.
- Applying the available forms of the nutrients at the right time due to calibrated
  rate as determined by the soil or leaf analysis test results.

Table 5.2 Examples of direct and indirect energy efficiency improvements in agriculture and fisheries through technical and social intervention (FAO/USAID 2015; Fang 2013)

Direct energy	Indirect energy
Fuel-efficient tractor engines/better maintenance	Less input-demanding crop varieties and animal breeds
More precise irrigation water application	Agro-ecological farming practices and nutrient recycling
Precision farming for accurate fertilizer application	Reducing water demand and losses
Adopting minimum or no-tillage practices	Improved fertilizer and machinery manufacture
Better control for building environments	GPS identification of fish stock locations to reduce trawling distances
Improved heat management of greenhouses	
Better propeller designs of fishing vessels	

- Applying frequent smaller amounts is more useful to the plant and crops more than adding the whole amounts in one dose.
- Applying liquid and soluble fertilizers, through injection, directly into irrigation system (fertigation).
- Using organic manures where available in line with good agricultural management as a slow release nutrient has a good effect on the soil aggregation, to increase water and nutrient retention (Table 5.2).

Chapter 6
Supplying Energy in a More Sustainable Way

6.1 Priority of Sustainability

A top priority now for the world is to make it more sustainable and accessible to all rural areas. To do this, we need to apply low-carbon emission renewable energy solutions to agriculture which is known as energy-smart food production to replace fossil fuels with its high gas emission. This increase is taking place in the heating, cooling and power sectors, and to some degree in the transport sector, especially in Europe and North America (through the growing use of biofuels and electric vehicles). In most rural areas especially in Africa and Asia, where no electricity grid connection exists, standalone mini-grid solutions are increasingly being constructed, particularly where they offer the potential to boost local economic development because of more intensive agricultural and food processing activities. Nowadays, a range of energy technologies and practices are accessible to many food chains, which provide opportunities to increase the access to modern energy and/or reduce fossil fuel demand—two intertwined objectives.

These include both renewable energy and energy efficiency measures, such as those illustrated below (FAO/USAID 2015).

Storage and refrigeration: Cooling and cold storages are used widely to maintain food quality both after harvesting and processing and to reduce losses along the supply chain. Refrigeration systems depend on reliable electricity supply systems, although new technologies such as solar absorption chillers are reaching the market. Other sources of renewable electricity can be used on both small and large scales. For cold stores, reducing energy demand is possible through such measures as increasing the insulation, keeping access doors closed and minimizing the heat load at the end of the processing phase of the cold chain (FAO/USAID 2015).

© The Author(s), under exclusive license to Springer Nature Switzerland AG 2020
N. Noureldeen Mohamed, *Energy in Agriculture Under Climate Change*,
SpringerBriefs in Climate Studies,
https://doi.org/10.1007/978-3-030-38010-6_6

6.2   Transport and Distribution

Under the fluctuating prices for fossil fuels, transport and distribution are particularly vulnerable and continue to raise components of the food chain. The key elements in this category are distance to the markets and the fuel prices in each specific country. Air-freighting for fresh food across the world to meet the demand for out-of-season products, and also sea freighting for cereal and grains is highly energy-dependent compared with supplying local markets with fresh and essential food when available. Transport of food commodities (such as milk powder or rice, wheat and other grain in bulk), fruit and vegetables (such as apples, bananas, potatoes and carrots), at times under controlled atmosphere or refrigeration, can be relatively cheap with a low-carbon footprint per ton. In rural areas, better roads can help reduce the energy and time needed to take fresh products to markets and hence improve local livelihoods (FAO/USAID 2015).

6.3   Field Machinery

Tractors and machinery can produce similar power outputs using less fuel where engines are maintained, tire pressures are correct, unnecessary ballast for the task is removed and the operator understands how to optimize tractor performance through correct gear and throttle selection as well as the use of the hydraulic systems. The suitable moisture soil contents are important to reduce fuel consumption, because very dry soil is more difficult in consuming much fuels without useful to soil tillage performance, and the same thing happens with high moisture content which causes tires slip with bad soil conditions after tillage needed to repeat the operation again. A well-trained operator can save up to 10% fuel and 20% of time sitting on the tractor, as well as reduce damage to both soil and machine through compacted solid or wheel-slip.

6.4   Food Processing

Processing of food at either the small-to-medium enterprise or large business scale requires energy for all operations such as heating, cooling, lighting, packaging and storing. The energy needed for such "beyond the farm gate" operations globally totals around three times the energy used "behind the farm gate" (Library of Congress Cataloging In Publication Data: Foth 1990; FAO/USAID 2015). In many processing crops, an energy audit by a trained specialist would identify cost-effective opportunities to reduce energy consumption while increasing throughput and quality of the products. Heat (such as for hot water, pasteurized milk, greenhouses, drying fruits and vegetables, and canned food) is normally produced by combusting natural gas, coal, oil and biomass, or from electrical heaters. In all cases, the heat can be provided from solar thermal, geothermal or modern bioenergy heat plants, or from efficiently designed heat pumps.

6.5 Renewable Energy in Agriculture

This type of energy can substitute fossil fuel inputs for heat and electricity all along the value-added chain where good local resources exist and government subsidies. It can be achieved using grid electricity with a growing share of renewables, or installing solar thermo-photovoltaic (PV), wind power or bioenergy for heat and power on the farm or at the processing plants. Since organic wastes are often produced both on-farm and at the processing plant, investments in anaerobic digestion plants to produce biogas that can be used to provide houses from heat, power or transport fuels are being widely deployed.

6.6 Fishing

The fishing industry can become more energy-smart along the entire food chain, particularly by reducing fuel consumption of large and small fishing vessels. This will help the industry cope with the volatility and rising trends of fuel and energy prices and ensure fish remain available at affordable prices for customers. For example, fouling (marine weed growth on the hull of a fishing vessel) can contribute to an increase in fuel consumption of up to 7% after only one month and 44% after 6 months, but it can be reduced significantly through the use of anti-fouling paints (FAO/USAID 2015). In addition, reducing 20% of the speed in a fishing vessel could reduce up to 51% of fuel consumption. It is worth pointing out that the development of energy-smart technologies can often lead to tradeoffs, including the following that needs to think about to find a solution that integrate between less energy uses, high yield crops and soil health:

- Between energy efficiency and efficiency in the use of other inputs (for instance, flood irrigation requires much less energy but is much less water efficient than drip irrigation).
- Between energy efficiency and access to energy (for instance, expensive energy-efficient tractors versus second-hand (most farmers in Middle-East used to make tillage in three directions: vertical, horizontal and radial in each tillage. This should be reduced to only one or two direction to save up to 50% of fuel and time), more affordable but much less energy-efficient ones; efficient biogas cook stoves versus woodfuel ones).

6.7   Energy Derived from the Agro-Food Chain

So far, we have stressed that energy is a key factor affecting agriculture and that potential options should be considered to enhance the sustainability of agricultural production and supply. However, agriculture, like other biomass-related economic sectors, offers the potential to produce biomass-based fuels. These biofuels can generate energy—called bioenergy or internal energy (self-sufficient) to supply various stages of the agricultural value chain, produce energy for external users and, once exported, generate additional income for the agriculture sector and producers. As such, they can initially supplement, and potentially substitute for, fossil fuels used in activities like transport, heating and cooking, and rural and industrial electrification (Fang 2013). Biofuels can come in three forms and types (FAO 2004):

- *Gaseous biofuels* (biogas and syngas) are produced from agricultural residues, woody residues or dedicated plantations through anaerobic digestion, gasification processes. Additional purification stages allow obtaining bio-hydrogen and bio-methane. These biofuels can potentially replace fossil fuels, such as natural gas, LPG and heating oils.
- *Liquid biofuels* (methanol, ethanol, butanol, biodiesel, bio-oil and straight vegetable oil) are produced from crops and other feedstock types (including biomass residues and algae). They can potentially replace fossil fuel alternatives such as diesel, petrol, propane and LPG. The most common alternatives for liquid biofuel production are fermentation, transesterification and pyrolysis or Fischer–Tropsch processes.
- *Solid biofuels* (charcoal, briquettes and pellets) comprise a transformation of biomass (like woody biomass and crop residues) into more fuel-efficient options through densification or pyrolysis processes. They can be used as intermediates for more valuable biofuels production or the potential replacement of fossil coals, LPG and propane. They can also substitute for fuelwood, especially in areas with deforestation problems.

Chapter 7
Conclusion

It is well known that 15% of global energy comes from water sources and 17% of energy is consumed by water lifting, desalination, treatment, distribution and delivery systems.

Energy in agriculture includes storage and refrigeration: Cooling and cold storage are used widely to maintain food quality both after harvesting and processing and to reduce losses along the supply chain. Refrigeration systems depend on reliable electricity supply systems, although new technologies such as solar absorption chillers are reaching the market. Other sources of renewable electricity can be used on both small and large scales. For cold stores, reducing energy demand is possible through such measures as increasing the insulation, keeping access doors closed and minimizing the heat load at the end of the processing phase of the cold chain. Energy for water includes water collection, processing, distribution and end-use power requirements, which are pumping, transport, treatment and desalination. Water for energy includes thermoelectric cooling, hydropower, mineral extraction and mining, fuel production (fossil, non-fossil and biofuel) and gas emission control.

Agriculture and energy including agriculture production (food, forage, fiber, and pharma and medicine plants) consume 30% of the world's available energy—with more than 70% consumed beyond the farm gate. The agriculture production produces about 31% of the world's greenhouse gas emissions. Almost one-third of the food we produce is lost or wasted. This, accompanying with another loss in energy, reached about 38% of the energy consumed in the agro-production chain. Developed countries use about 35 GJ per person a year for food and agriculture (nearly half in processing and distribution). Developing countries use only 8 GJ per person per year (nearly half for cooking).

The relation between climate, water, soil and crop production is so complicated and looks like a circle without a beginning or an end, whereas each item affects all others. Global heating will increase the evaporation from all waterbodies (rivers, irrigation canals, lakes etc.) causing a lack of water especially in hyper arid, arid and semi-arid regions. At the same time, it will decrease the amounts of little

© The Author(s), under exclusive license to Springer Nature Switzerland AG 2020
N. Noureldeen Mohamed, *Energy in Agriculture Under Climate Change*,
SpringerBriefs in Climate Studies,
https://doi.org/10.1007/978-3-030-38010-6_7

rainfall. Global heating also will increase the evaporation and evapotranspiration from irrigated lands, through grasses, weeds and growing plants. Thus, the plants will need more water and more frequent irrigation, and the soil also will need more water due to the salinity building up and salt accumulation. But actually there is a lack of water in addition to deterioration of water quality due to the increase of pollutants concentration because of high temperature effects, which will also increase land degradation and lead to the need for more water requirement at the time of water scarcity. Water scarcity in hyper arid and arid countries will lead to expansion in reusing of wastewater such as agriculture drainage water which has appreciable amounts of salts, residual chemical fertilizers and residual pesticides, which are classified as problematic water and will cause more land degradation. Thus, this poor quality water will need using of more water for leaching as leaching fraction or leaching requirements, but the sufficient water to meet all these demands is completely absent. Finally, climate change will decrease available water and will increase land degradation and building up of salt-affected soils and contaminated soils, and this will decrease land productivity which needs sufficient water to leach out the salinity and contaminators, but there is not enough water to meet these demands!!. The following drawing diagram summarizes the relation between global heating and climate change.

The correlation and integration between water, energy and food (WEF) comes true and becomes so close. We will never use again the terms such as "the future of food production", or the "future of energy demands", or the future of water security. These three have become one nexus that includes water, energy and food. Moreover, with the climate change reality, the nexus accepts it as the fourth item and nearly will accept also the population growth. Thus starting from the year 2018 the world uses the terms the "future of nexus of water, energy, food, and climate change and population growth". This is logic because no agriculture productions without energy, no food without water and no energy without water and food, and the global population will be 9.6 billion capita in year 2050 instead of 7.6 billion in 2018. Thus we are talking about one integrated nexus with no possibility to separate any of its items.

The energy consumption to produce the major nutrients as chemical fertilizers of nitrogen, phosphate and potassium reflects that the nitrogen production is heavier energy consumer than phosphate and potassium and is represented eight-fold (time). Moreover, there is energy consumption in producing pesticides hormones and growth regulators. The following figures show the correlation between energy and agriculture in producing fertilizers and pesticide in addition to increase in demand on electricity periodically and the energy to and from the food value chain.

With global heating, FAO estimated that the food productions should be increased by 50% before 2050 from only 20 to 30% more from water. This means the amount of agricultural chemical such as fertilizers, pesticides and growth regulators and hormones will increase to meet the need of the increasing food and other agriculture production by 60%. The climate condition will cause a decrease of available water by 6% with lowering of its quality, may increase the energy consumption to be doubled of it before year 2100 and by 50% by year 2030, as shown in Figs. (5.1, 5.2, 5.3 and 5.4). Thus the traditional nexus of water, energy, food is modified to include climate change (water, food, energy and climate change) and we suggested in last meeting of FAO in June 2018, for Soil Partnership Plenary assemble to modify it again to include population growth (water, food, energy, climate change and population growth).

The top priority now for the world is to make it more sustainable and accessible to all rural areas. To do this, we need to apply low-carbon emission renewable energy solutions to agriculture which is known as energy-smart food production to replace fossil fuels with its high gas emission. This increase takes place in the heating, cooling and power sectors, and to some degree in the transport sector, especially in Europe and North America (through the growing use of biofuels and electric vehicles). In most rural areas especially in Africa and Asia, where no electricity grid connection exists, standalone mini-grid solutions are increasingly being constructed, particularly where they offer the potential to boost local economic development because of more intensive agricultural and food processing activities. Nowadays, a range of energy technologies and practices are accessible to many food chains, which provide opportunities to increase the access to modern energy and/or reduce fossil fuel demand—two intertwined objectives.

These include both renewable energy and energy efficiency measures, such as storage and refrigeration: Cooling and cold storage are used widely to maintain food quality, both after harvesting and processing, and to reduce losses along the supply chain. Refrigeration systems depend on reliable electricity supply systems, although new technologies such as solar absorption chillers are reaching the market. Other sources of renewable electricity can be used on both small and large scales. For cold stores, reducing energy demand is possible through such measures as increasing the insulation, keeping access doors closed and minimizing the heat load at the end of the processing phase of the cold chain (FAO/USAID 2015).

In the year 2050 the demands for water will increase by 30%, but under global heating the total water resources will decrease by 6%. Moreover, food production will decrease by 12–20% and land degradation will increase according to building up of soil salinity and land degradation due to increase of evaporation. Increase of soil salinity means the need for much water will increase to achieve salinity leaching out. Decrease of the accessible water will increase the needed for treated wastewater for reuse, desalination of sea water and even treated the marginal well water that contains appreciable amounts of salt.

Drought and dryness frequencies will increase under global heating especially in upstream river basins, thus the need for more groundwater and desalination water will increase sharply and the energy consumption will increase.

The need for more food production will increase by 60% in 2050 which means the need for more chemical fertilizers, pesticides and fish will also increase, which is considered as high intensive energy consumption. Water scarcity under global heating will increase the need for intensive food production under greenhouse systems in addition to the need for air-condition and fast transportation for the productive foods will also increase the energy consumption.

References

ADB (2009) Improving energy security and reducing carbon intensity in Asia and the Pacific. Asian Development Bank, Manila, Philippines, p 48

Alexandratos N, Bruinsma J (2012) World agriculture towards 2030/2050: the 2012 revision. ESA Working paper No. 12-03, Food and Agriculture Organization of the United Nations (FAO), Rome, Italy.

Barghouti S (2009) Climate change in the MENA region: effects on water and agriculture; addressing climate change in the Middle East and North Africa (MENA) regions launch conference, 13–14 May 2009, Rome, Italy

Bundschuh J (2014) Sustainable energy solutions in agriculture.CRC Press/Balkema is an imprint of the Taylor & Francis Group, an informa business © 2014 Taylor & Francis Group, London, UK

Bundschuh J, University of Southern Queensland (USQ) (2014) Toowoomba, Australia Royal Institute of Technology (KTH), Stockholm, Sweden 2014. Sustainable Energy Developments, vol 8, ISSN: 2164-0645

CAE (1996) Energy efficiency—a guide to current and emerging technologies, vol 2, Industry and Primary Production. Centre for Advanced Engineering, University of Canterbury, Christchurch, New Zealand

Comprehensive Assessment (2007) Trends in water and agricultural development. Integrated Water Management Institute, India. http://www.iwmi.cgiar.org/assessment/water%20for%20food%20water%20for%20life/chapters/chapter%202%20trends.pdf

Cooley H, Gleick P, Wolff G (2006) Desalination, with a grain of salt. Pacific Institute, Oakland

CGIAR Research Program on Climate Change, Agriculture and Food Security (CCAFS) (n.d.) Food security. https://ccafs.cgiar.org/bigfacts/#theme=food-security

Edame GE, Ekpenyong AB, Fonta WM, Duru EJC (2011) Climate change, food security and agricultural productivity in africa: issues and policy directions

Energy Information Administration (EIA) 2016 Monthly energy review, Sept 2016. Energy Information Administration, U.S. Department of Energy, Washington, DC

Environmental Protection Agency (EPA) (2015) Climate change in the United States: benefits of global action. Office of Atmospheric Programs, Environmental Protection Agency, Washington, DC

ET Energy World (2019) World's top 10 countries in wind energy capacity, Mar 2019. https://energy.economictimes.indiatimes.com/news/renewable/innovation-kochis-cift-launches-solar-fish-dryer/70507417. Accessed 3 Aug 2019

© The Author(s), under exclusive license to Springer Nature Switzerland AG 2020   69
N. Noureldeen Mohamed, *Energy in Agriculture Under Climate Change*,
SpringerBriefs in Climate Studies,
https://doi.org/10.1007/978-3-030-38010-6

European Report on Development (2012) Confronting scarcity: managing water, energy and land for inclusive and sustainable growth. https://ec.europa.eu/europeaid/sites/devco/files/erd-consca-report-20110101_en_0.pdf

Energy security define as the uninterrupted of energy sources at an affordable price [IEA].

Fang Z (2013) Liquid, gaseous and solid biofuels—conversion techniques. In Tech.

FAO (2004) United bioenergy terminology. Rome, Italy Food and Agriculture Organization of the United Nations, Rome. ftp://ftp.fao.org/docrep/fao/007/j4504e/j4504e00.pdf

FAO (2006) Food security, policy brief, June 2006, Issue 2. http://www.fao.org/forestry/13128-0e6f36f27e0091055bec28ebe830f46b3.pdf

FAO (2007) Feed supplementation of blocks. In: Animal production and health paper 164. Food and Agriculture Organization of the United Nations, Rome. http://www.fao.org/docrep/010/a0242e/a0242e00.htm) .

FAO (2011) The state of world fisheries and aquaculture 2010. Food and Agriculture Organization of the United Nations, Rome

FAO (2012) Energy-smart food at FAO: an overview. FAO, Viale delle Terme di Caracalla, 00153 Rome, Italy.

FAO (2016) Reports about the effects of climate change, agriculture and food security stated that:

FAO (Food and Agriculture Organization of the United Nations) (2010) How to feed the world in 2050. Food and Agriculture Organization of the United Nations, Rome, Italy, 2009. www.fao.org/fileadmin/templates/wsfs/docs/expert_paper/How_to_Feed_the_World_in_2050.pdf. Accessed June 2019

FAO (Food and Agriculture Organization of the United Nations) (2011) Energy-smart food for people and climate. Issue Paper, Food and Agriculture Organization of the United Nations, Rome, Italy, 2011. http://www.fao.org/docrep/014/i2454e/i2454e00.pdf. Accessed May 2019

FAO/USAID (2015) Opportunities for agri-food chains to become energy-smart. http://www.fao.org/publications/card/en/c/0ca1c73e-18ab-4dba-81b0-f8e480c37113/

Food and Agriculture Organization (FAO) (2000a) The energy and agriculture nexus, environment and natural resources Working Paper No. 4, FAO, Rome, 2000. http://www.fao.org/docrep/014/i2454e/i2454e00.pdf. Accessed Aug 2019

Food and Agriculture Organization (FAO) (2000b) The state of food insecurity in the world, 1999, FAO press summary, 2000, Viale delle Terme di Caracalla, 00153 Rome, Italy.

Food and Agriculture Organization of the United Nations (FAO) (2000) FAO/ESAF handbook for defining and setting up a Food Security Information and Early Warning System (FSIEWS). Rome

Food and Agriculture Organization of the United Nations (FAO) (2011a) Energy-smart food for people and climate, Viale delle Terme di Caracalla, 00153 Rome, Italy.

Food and Agriculture Organization of the United Nations (FAO) (2011b) The State of the World's Land and Water Resources for Food and Agriculture (SOLAW). Food and Agriculture Organization of the United Nations, Rome.

Food and Agriculture Organization of the United Nations (FAO) (2012) Energy-smart food for people and climate: an overview. FAO, Rome, Italy

Food and Agriculture Organization of the United Nations Rome (2014) The water-energy-food nexus a new approach in support of food security and sustainable agriculture.

GEMCO Energy (2011) Top 10 countries interested in biomass energy. http://www.biomass-energy.org/blog/top-10-countries-interested-in-biomass-energy.html. Accessed 25 July 2019

GEA (2012) Global energy assessment—towards a sustainable future. IIASA, Vienna, Austria. http://www.iiasa.ac.at/web/home/research/researchPrograms/Energy/Home-GEA.en.html. Accessed May 2019.

Hoff H (2011) Understanding the nexus (Background paper for the Bonn 2011) http://www.fao.org/docrep/014/i2454e/i2454e00.pdf. Accessed Aug 2018

Hydraulic Institute, Europump, and The U.S. Department of Energy (2004) Variable speed pumping—a guide to successful applications, executive summary. Hydraulic Institute, Europump, and the U.S. Department of Energy's (DOE) Industrial Technologies Program

Irfan U (2019) America's record high energy consumption, explained in 3 charts (https://www.vox.com/2019/4/18/18412380/climate-change-2019-energy-natural-gas,, Apr 18, 2019.

IEA (2007) Energy security and climate policy: assessing interactions. International Energy Agency, IEA/OECD, Paris, France, 2007

IEA (2011) World energy outlook 2011. International Energy Agency, IEA/OECD, Paris, France, p 72

IISD REPORT FEBRUARY© 2 2001313 The International Institute for Sustainable Development The Water–Energy–Food Security Nexus: towards a practical planning and decision-support framework for landscape investment and risk management.

International Energy Agency (IEA) (2014) Energy, climate change and environment: 2014 insights, OECD/IEA, 2014 International Energy Agency 9 rue de la Fédération 75739 Paris Cedex 15, France. www.iea.org

Int J Humanit Soc Sci 1(21):205 (Special Issue—Dec 2011)

IPCC (2007a) Summary for policymakers. In: Climate change 2007: impacts, adaptation and vulnerability. In: Parry ML, Canziani OF, Palutikof JP, van der Linden PJ, Hanson CE (eds) Contribution of Working Group II to the fourth assessment report of the intergovernmental panel on climate change, Cambridge University Press, Cambridge, UK, pp 7–22

IPCC (2007b) Freshwater resources and their management. In: Kundzewicz ZW, Mata LJ, Arnell NW, Döll P, Kabat P, Jiménez B, Miller KA, Oki T, Sen Z, Shiklomanov IA. Climate change 2007: impacts, adaptation and vulnerability. contribution of Working Group II to the fourth assessment report of the intergovernmental panel on climate change, Parry ML, Canziani OF, Palutikof JP, van der Linden PJ, Hanson CE (eds). Cambridge University Press, Cambridge, UK, pp 173–210.

IPCC (2007) Fourth assessment report. Intergovernmental panel on climate change. Cambridge University Press, Cambridge.

International Energy Agency (IEA) (n.d.) Energy Security. https://www.iea.org/topics/energysecurity/

Jones MJJ (2012) Thematic Paper 8: social adoption of groundwater pumping technology and the development of groundwater cultures: governance at the point of abstraction. Groundwater governance: a global framework for country action. GEF ID 3726. Food and Agriculture Organization of the United Nations Rome

Library of Congress Cataloging In Publication Data: Foth HD (1990) Fundamentals of soil science/Foth HD 8th edn. p. cm. ISBN 0-471-52279-1 1. Soil science. I. Title. S591.F68 1990, 631.4-dc20

Mohcine Bakhat a,b, Klaas Würzburg (2013). Price relationships of crude oil and food commodities, economic of energy WP FA06/2013.

Müller C, Bondeau A, Popp A, Waha K, Fader M (2009) Climate change impacts on agricultural yields. Background note for the WDR 2010

Mundi I (2012) Historical commodity prices. http://www.indexmundi.com/commodities. Accessed June 2019

Murray M (2008) 'The water energy nexus'. In: Power point presentation. Donald Bren School of Environmental Science and Management, University of California Santa Barbara, 8–9 May 2008

Mushtaq S, Maraseni NT (2011) Technological change in the Australian irrigation industry: implications for future resource management and policy development. Waterlines Report Series No [53.], Aug 2011, National Water Commission, Canberra, ACT, Australia, 2011. http://archive.nwc.gov.au/__data/assets/pdf_file/0013/10921/Waterlines_53_PDF_Fellowship-_Technological_change_in_the_irrigation_industry.pdf. Accessed July 2019

NASA (2011) Earth's average temperature is expected to rise even if the amount of greenhouse gases in the atmosphere decreases. But the rise would be less than if greenhouse gas amounts remain the same or increase. https://www.nasa.gov/audience/forstudents/5-8/features/nasa-knows/what-is-climate-change-58.html

Nexus Regional Dialogue Programme (NRD) (2018) NEXUS water-energy-food dialogues train-ing material training Unit 01: introduction to the Water-Energy-Food Security (WEF) NEXUS. Authors Ian McNamara Alexandra Nauditt Santiago Penedo Lars Ribbe TH Köln—University of Applied Sciences Institute for Technology and Resources Management in the Tropics and Subtropics (ITT) Betzdorfer Str. 2 D- 50679 Köln, Germany

Oxfam Australia (2012) What's wrong with our food system?. https://www.oxfam.org.au/2012/05/whats-wrong-with-our-food-system/

Pimentel D, Pimentel MH (2009) Ecological systems, natural resources, and food supplies. In: Pimentel D, Pimentel MH (eds) Food, energy, and society. 3rd edn. CRC Press of the Taylor & Francis Group, Boca Raton, FL, 2009, pp 21–34

Plant Growth Regulators (2012) https://ar-ar.facebook.com/notes/%D9%85%D9%85%D9%84%D9%83%D8%A9-%D9%86%D8%A8%D8%A7%D8%AA%D8%A7%D8%AA-%D8%A7%D9%84%D8%B2%D9%8A%D9%86%D8%A9/%D9%85%D9%86%D8%B8%D9%85%D8%A7%D8%AA-%D8%A7%D9%84%D9%86%D9%85%D9%88-%D8%A7%D9%84%D9%86%D8%A8%D8%A7%D8%AA%D9%8A-plant-growth-regulators-r%C3%A9gulateurs-de-croissance/365353736888637/

Pinterest site. https://www.pinterest.co.uk/pin/717901996820657072/. Accessed 22 Mar 2018

REN21 (2012) Renewables 2012: Global status report, Paris.

Scientific American E, News E (2019) How Co-ops are bringing solar power to rural America Mar 2019. https://www.scientificamerican.com/article/how-co-ops-are-bringing-solar-power-to-rural-america/. Accessed Aug 2019

Steenblik R (2007) 'Biofuels—at what cost? government support for ethanol and biodiesel in selected OECD countries. International Institute for Sustainable Development, Geneva. http://www.iisd.org/pdf/2007/biofuels_oecd_synthesis_report.pdf

Suuronen P, Chopin F, Glass C, Løkkeborg S, Matsushita Y, Queirolo D, Rihan D, forthcoming (2012) Low impact and fuel efficient fishing-looking beyond the horizon. Fisheries Research

The Hamilton Project and Energy Policy Institute at the University of Chicago (2017) Twelve economic facts on energy and climate change, The Hamilton Project Energy Policy Institute at the University of Chicago, Mar 2017

United Nations Department of Economic and Social Affairs (UNDESA) (2014) International decade for action WATER FOR LIFE 2005–2015. http://www.un.org/waterforlifedecade/water_and_energy.shtml

United nations Development Programme (UNDP) (2007) Human development report 2007/2008. fighting climate change: human solidarity in a divided world. New York

United Nations Environment Programme (UNEP) (2012) A glass half empty: regions at risk due to groundwater depletion why is this issue important? https://na.unep.net/geas/getUNEPPageWithArticleIDScript.php?article_id=76

United Nations University (UNU) (2013) Water security & the global water agenda: A UN-Water Analytical Brief, Hamilton.

U.S. Department of Agriculture, Economic Research Service (ERS) (2011) Cost of Production Database

U.S. Department of Agriculture, Economic Research Service (ERS) (2012) Monthly energy review february 2012. Office of Energy Statistics

US Energy Information Administration (2019) In 2018, the United States consumed more energy than ever before, 16 Apr 2019

U.S. Energy Information Administration (EIA) (2017) International energy outlook 2017. https://www.eia.gov/outlooks/ieo/pdf/0484(2017).pdf

Verma SR (2008) Impact of agricultural mechanization on production, productivity, cropping inten-sity income generation and employment of labour. Published in the Status of Farm Mechaniza-tion in India, Indian Agricultural Statistics Research Institute, New Delhi. http://agricoop.nic.in/Farm%20Mech.%20PDF/05024-03.pdf)

WATERSHED ACADEMY WEB and US Environmental Protection Agency (2013). The effect of climate change on water resources and programs. https://cfpub.epa.gov/watertrain/pdf/modules/Climate_Change_Module.pdf. Accessed 3 July 2019

World Bank (2000) World development indicators. World Bank, Washington, p 2000

World Bank (2008) World development indicators 2008. World Bank, Washington, DC

World Bank (2010) World development indicators. The World Bank Group, Washington. http://data.worldbank.org/data-catalog/world-development-indicators/wdi-2010

World Economic Forum WEF (2011) Global risks 2011, 6th edn. World Economic Forum, Cologne/Geneva

World Energy Council (2016) World energy resources: bioenergy 2016. https://www.worldenergy.org/wp-content/uploads/2017/03/WEResources_Bioenergy_2016.pdf

Zentner RP, Lafond GP, Derksen DA, Nagy CN, Wall DD, May WE (2004) Effects of tillage method and crop rotation on non-renewable energy use efficiency for a thin Black Chernozem in the Canadian Prairies. Soil Till Res 77(2004):125–136

Printed in the United States
By Bookmasters

Printed in the United States
By Bookmasters